Reviews of Physiology, Biochemistry and Pharmacology 157

Reviews of

157 Physiology Biochemistry and Pharmacology

With 16 Figures and 5 Tables

Springer

Library of Congress-Catalog-Card Number 74-3674

ISSN 030-4240

ISBN 978-3-540-39688-8 Springer Berlin Heidelberg New York

Springer is a part of Springer Science+Business Media

springer.com

Editor: Simon Rallison, London
Desk Editor: Anne Clauss, Heidelberg
Production Editor: Patrick Waltemate, Leipzig
Typesetting and Production: LE-TEX Jelonek, Schmidt & Vöckler GbR, Leipzig
Cover: WMXDesign GmbH, Heidelberg

Printed on acid-free paper 14/3100YL – 5 4 3 2 1 0

Rev Physiol Biochem Pharmacol (2006)
DOI 10.1007/112_0508

E. Bettiol · S. Clement · K. H. Krause · M. E. Jaconi

Embryonic and adult stem cell-derived cardiomyocytes: lessons from in vitro models

Published online: 29 April 2006

Abstract For years, research has focused on how to treat heart failure by sustaining the overloaded remaining cardiomyocytes. Recently, the concept of cell replacement therapy as a treatment of heart diseases has opened a new area of investigation. In vitro-generated cardiomyocytes could be injected into the heart to rescue the function of a damaged myocardium. Embryonic and/or adult stem cells could provide cardiac cells for this purpose. Knowledge of fundamental cardiac differentiation mechanisms unraveled by studies on animal models has been improved using in vitro models of cardiogenesis such as mouse embryonal carcinoma cells, mouse embryonic stem cells and, recently, human embryonic stem cells. On the other hand, studies suggesting the existence of cardiac stem cells and the potential of adult stem cells from bone marrow or skeletal muscle to differentiate toward unexpected phenotypes raise hope and questions about their potential use for cardiac cell therapy. In this review, we compare the specificities of embryonic vs adult stem cell populations regarding their cardiac differentiation potential, and we give an overview of what in vitro models have taught us about cardiogenesis.

Introduction

The heart is a fascinating organ that has always aroused great interest. Traditionally believed to be the location of feelings, it has today become the location of one of the most common pathologies of our time and, therefore, one of the biggest challenges for modern medicine.

E. Bettiol · S. Clement · M. E. Jaconi (✉)
University of Geneva, Department of Pathology and Immunology, Faculty of Medicine, Geneva, Switzerland
e-mail: Marisa.Jaconi@medecine.unige.ch · Tel.: +41-22-3795257 · Fax: +41-22-3795746

K. H. Krause
Geneva University Hospitals, Biology of Aging Laboratory, Department of Geriatrics and Rehabilitation,
Geneva, Switzerland

Cardiovascular diseases are a leading cause of morbidity and mortality both in developed and developing countries. Therefore, understanding cardiac developmental and physiopathological mechanisms has become indispensable, not only for fundamental research purposes, but also because modern medicine has a heavy burden in its task to find new and more efficient ways to cure diseased hearts. Indeed, upon major injury, the heart is unable to regenerate itself, unlike other organs such as liver, skin, or bone. Studies of mammalian fetal and adult hearts or isolated cardiomyocytes have proven essential to establish pharmacological and surgical treatments. However, these treatments only act on the overloaded remaining cardiomyocytes and, ultimately, terminal heart failure ensues.

Regenerative medicine, or more specifically cardiac cell therapy, profiles itself as a possible way to rescue a damaged heart by implanting new cells to replace the damaged cardiomyocytes and thus regenerate the cardiac muscle. Proof of concept for the engraftment of cells in the myocardium has been provided, but definitive demonstration of cell functionality, i.e., electrical coupling and contractile performance, still needs to be confirmed. The best source of cells is not clear yet and in vitro and in vivo experiments are on the way to answering these questions. In vitro-generated cardiomyocytes are a promising source, and therefore a thorough knowledge of fundamental cardiac differentiation mechanisms is mandatory.

In this review, we focus on what we have learned of cardiac differentiation from in vitro models using embryonic and adult stem cells.

Cardiac differentiation

In vivo, cardiomyocytes derive from the lateral plate mesoderm. Studies conducted on Drosophila, mice, amphibians, and avian embryos have shed light on some of the mechanisms leading to cardiac induction. The role of pro- or anti-cardiogenic growth factors and morphogens is crucial in this process and Fig. 1 summarizes results obtained on these models. Indeed, secretion of these factors during embryogenesis by cells surrounding the cardiogenic mesoderm activates signaling cascades that, through the expression of cardiac transcription factors, finally leads to a functional cardiac phenotype (for review see Brand 2003; Fishman and Chien 1997; Srivastava and Olson 2000). Identifying the complex interplay of growth factors, their underlying signaling pathways and target genes is of great importance to enable the specific differentiation of stem cells toward the cardiac phenotype.

Crucial factors with pro-cardiogenic effects are secreted by the endoderm (Lough and Sugi 2000), such as several members of the transforming growth factor β (TGFβ) superfamily. For instance, treatment of *Xenopus Laevis* embryo explants with activin A can induce cardiac differentiation (Logan and Mohun 1993) and bone morphogenetic protein-2 (BMP-2) and fibroblast growth factor-4 (FGF-4) together can trigger cardiogenesis in nonprecardiac mesoderm of chicken embryos (Lough et al. 1996). On the other hand, Cripto, a growth factor expressed in early mouse mesoderm (Dono et al. 1993), has a permissive effect on cardiac differentiation by allowing Nodal, another member of the TGFβ superfamily, to bind to its receptor (Schier and Shen 2000).

The Wnt/β-catenin pathway also plays an important role in vivo. The canonical Wnt pathway (Wnt1, 3, 8) that ultimately activates β-catenin has an anti-cardiogenic effect. Indeed, conditional knockout of the β-catenin gene in the definitive endoderm induced an ectopic expression of BMP-2 and the formation of multiple hearts (Lickert et al. 2002). Antagonists of the canonical Wnt pathway such as Dickkopf-1 or Crescent play an indirect role in heart induction through the transcription factor Hex (Foley and Mercola 2005). In opposi-

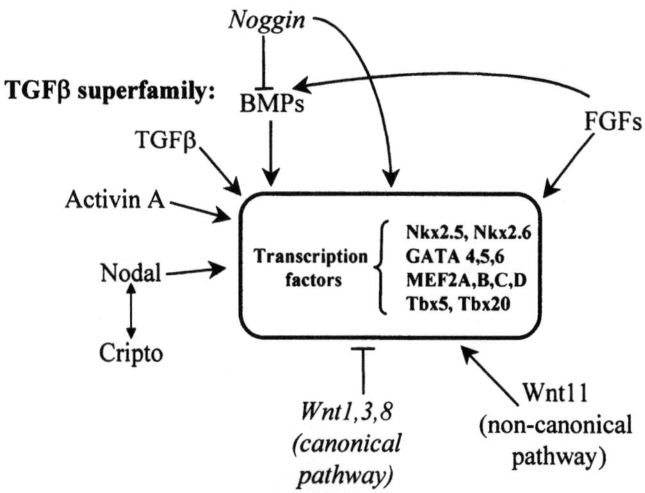

Fig. 1 General scheme of the different growth factor families involved in cardiac differentiation. Results summarized in this figure were obtained on *Drosophila*, mice, amphibians, or avian embryos. Activation of downstream pathways induces the expression of Nkx-, MEF2-, GATA-, and T-box types of transcription factors

tion, the noncanonical Wnt pathway, which activates protein kinase C (PKC)-dependent signaling, enhances cardiogenesis. Studies on avian embryos highlighted the need for Wnt11 (Eisenberg and Eisenberg 1999; Eisenberg et al. 1997) in cardiac differentiation, as it induced ectopic cardiogenesis.

In cardiac precursor cells, activating or inhibiting signals (growth factors, morphogens, peptides) control the expression of a complex network of transcription factors. The Nkx (Evans 1999), MEF2 (Black and Olson 1998), GATA (Molkentin 2000), and T-box families (Ryan and Chin 2003) of transcription factors play roles in cardiac differentiation. Members of different families can interact or affect each other's expression (Durocher et al. 1997; Morin et al. 2000). Several of them are mainly expressed in the heart, but none of them are exclusively expressed in cardiac cells.

Stem cells and stem cell-derived cardiomyocytes

Stem cells are defined by their ability to self-renew as well as to enter differentiation pathways. They are classified according to their differentiation potential. If they can give rise to cells of the three germ layers, they are defined as pluripotent. If they can differentiate toward a subset of cell types, they are called multipotent. When their differentiation potential is restricted to one cell type, they are unipotent.

Stem cells lines derived at an embryonic stage, such as embryonal carcinoma cells (ECCs) and embryonic stem cells (ESCs) are pluripotent. They are a powerful tool to study in vitro early developmental processes such as germ layer and lineage specification (for review see Gadue et al. 2005; Keller 2005). On the other hand, adult stem cells, located in many organs of the adult body and responsible for the turnover and healing of tissues, are usually multipotent or unipotent. Cardiomyocytes can derive from different types of adult or embryonic stem cells (Fig. 2). In this review, we will give a comprehensive overview of the different reports about the cardiogenic potential of both embryonic and adult stem

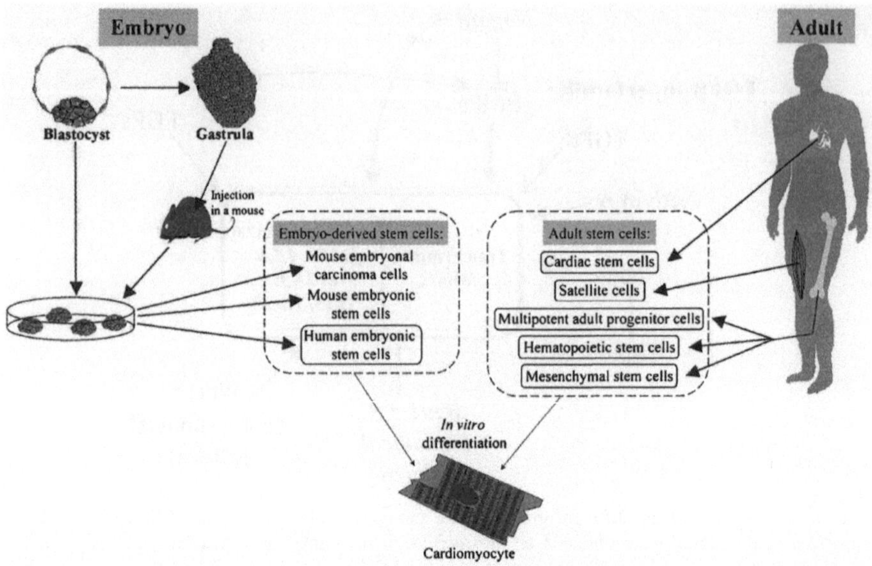

Fig. 2 Two types of stem cells can differentiate toward the cardiac phenotype in vitro: embryo-derived and adult stem cells. Cell types in *squares* could be used in the future for cell therapy

cells, such as hematopoietic stem cells, mesenchymal stem cells, multipotent adult progenitor cells, cardiac stem cells, and satellite cells.

Stem cells, as in vitro differentiation models, offer a great chance to learn more about fundamental differentiation mechanisms. Their study can not only give hope for cellular replacement therapies, but will also increase our knowledge in developmental and cell biology as well as molecular biology and physiopathology.

Cardiac differentiation of embryo-derived stem cells

Several models of embryo-derived stem cells are now available for the study of cardiac differentiation in vitro. Mouse embryonal carcinoma cells (mECCs), mouse embryonic stem cells (mESCs), and human embryonic stem cells (hESCs) are derived from early embryos and are pluripotent. Under specific conditions, cardiac differentiation occurs spontaneously and can therefore be studied in vitro. Transcription factors expression and interactions, myofibrillogenesis, electrophysiological characteristics, and embryonic vs neonatal or adult cellular features can be studied. These models are also useful to detect new players in cardiac differentiation and to enhance cardiogenesis.

Mouse embryonal carcinoma cells

Mouse embryonal carcinoma cells (mECCs) are the first model that has allowed the in vitro study of cardiac differentiation. Mouse ECCs constitute the stem cell population found in teratocarcinomas, a tumor deriving from primordial germ cells. Several multipotent and euploid cell lines were derived from teratocarcinomas generated by subcutaneous injection of mouse embryos undergoing gastrulation (Kleinsmith and Pierce 1964; Martin and Evans

1974). Mouse ECC cell lines grow easily in culture for infinite periods of time and are pluripotent, as they form teratomas when injected in nude mice. In vitro, mECC differentiation is induced by letting cells aggregate in suspension and form three dimensional (3D) structures named embryoid bodies (EBs) (Martin and Evans 1975).

Most studies on mECCs were conducted on the P19 cell line (McBurney and Rogers 1982). Cardiac differentiation can be induced by cell aggregation and addition of low concentrations of dimethyl-sulfoxide (DMSO) (Edwards et al. 1983; McBurney et al. 1982) or retinoic acid (RA) (Edwards and McBurney 1983). Using this protocol, some cells differentiate into cardiomyocytes and small clumps start to beat spontaneously. Another cell line differentiating with an efficiency of about 90% toward the cardiac phenotype was subcloned from P19 cells and named P19CL6 (Habara-Ohkubo 1996). However, without these chemical treatments, there is no spontaneous cardiac differentiation.

Characterization of mECC-derived cardiomyocytes

Mouse ECC-derived cardiomyocytes express cardiac sarcomeric proteins such as α-cardiac and α-skeletal actins, myosin heavy chain (MHC), myosin light chain (mlc) 2 atrial and mlc1 (Rudnicki et al. 1990). Action potentials (APs) and ion currents can be recorded in these cells and their properties change as cells mature (van der Heyden et al. 2003; Wobus et al. 1994). These studies indicate that the mECC model closely recapitulates in vivo mouse cardiogenesis. However, the cells remain at an embryonic or neonatal stage of development.

The mECC model has served for the identification of new transcriptional regulators of cardiogenesis and their target genes. For example, an early transcription factor named Myocytic Induction/Differentiation ORIginator (MIDORI) was identified using differential mRNA display (Hosoda et al. 2001). This protein is expressed in vivo at day 7.5 in the cardiac crescent of mouse embryos and its overexpression in P19 cells increased the number of myosin-positive cells. More complex gene expression studies using cDNA microarrays on P19CL6 cells confirmed the upregulation of several known genes, such as eHAND, MEF2C and mlc, but also highlighted changes in the expression levels of many expressed sequence tags (ESTs) whose function is still unknown and which could therefore be early regulators of cardiogenesis (Peng et al. 2002). Liu et al. (2005) investigated the effect of downregulation or inducible overexpression of Nkx2.5 on gene expression by cDNA microarrays. This study highlighted a positive transcriptional effect of Nkx2.5 on genes involved in TGFβ signaling and on Brachyury, a T-box transcription factor expressed in mesoderm, as well as a negative effect on Jumanji, a cardiac transcriptional repressor.

Mouse ECC: factors and pathways involved in cardiac differentiation

During cardiac differentiation, growth factors and morphogens secreted by other germ layers play a key role in cardiac induction or inhibition (Fig. 1). The requirement for a primitive extraembryonic endoderm cell layer around the cardiac cells was first demonstrated in EBs derived from mECCs (Smith et al. 1987). Later, through the coculture of P19 cells with a visceral endoderm-like cell line (END-2), Mummery et al. (1991) showed that only visceral endoderm- and not parietal endoderm- or mesodermal-like derivatives could induce cardiac differentiation in a DMSO-independent manner. In fact, END-2 supernatant was sufficient to activate the process, demonstrating the role of endoderm-derived soluble factors.

Bone morphogenetic proteins (BMPs) are members of the TGF-β superfamily of growth factors that are secreted by the endoderm. Their effect was investigated by creating a P19CL6 cell line overexpressing noggin, a BMP antagonist (Monzen et al. 1999). This cell line was unable to differentiate into cardiomyocytes, but the phenotype was rescued by overexpression of BMP-2, TAK1 (a mitogen activated protein kinase kinase kinase) or Nkx2.5 and GATA-4 together, as well as by adding exogenous BMPs to the culture medium. These results demonstrate the role of BMPs in cardiac induction and define the p38/MAPK pathway as one of the downstream signaling pathways, as confirmed by other studies (Davidson and Morange 2000; Eriksson and Leppa 2002). Smad proteins, known mediators in the TGF-β signaling pathway, have also been proposed as downstream effectors. Indeed, the co-expression of Smad1 and Smad4 restored cardiac differentiation in the noggin-overexpressing P19CL6 cells, through activation of the transcription factor ATF-2 (Monzen et al. 2001). Another downstream pathway involved in BMP-2 signaling has been implicated in cardiac differentiation of mECCs. Indeed, MEF-2A and MHC expression can be activated by BMP-2 in P19CL6 cells through a mechanism depending on phosphatidyl-inositol-3-kinase (PI-3-kinase) (Ghosh-Choudhury et al. 2003). Naito et al. (2003) showed that the inhibition of PI-3-kinase by Wortmannin or LY294002 could abolish cardiac differentiation. In summary, TGF-β and BMPs seem to favor cardiogenesis through activation of at least three pathways, two with ATF2 as a common target involving p38 MAPK and/or Smad proteins and one involving PI-3-kinase.

Studies on mECCs showed that activation of the noncanonical pathway by Wnt11 can regulate cardiogenesis (Pandur et al. 2002). Indeed, Wnt11-conditioned medium triggered cardiac differentiation in the absence of DMSO. However, Nakamura et al. (2003) reported that, in the same model, canonical Wnt proteins (Wnt3a and Wnt8a), known for their inhibitory effect on cardiogenesis, are nevertheless expressed as well and that β-catenin can be found in its activated form. In particular, Wnt3a-conditioned medium tripled the number of MHC-positive cells. The latter results do not fit with in vivo data, but this could be explained by the fact that differentiating P19CL6 cells are not a homogenous population. Some of the proteins studied could therefore be expressed in cardiac and/or noncardiac cells and might therefore have direct or indirect effects on cardiogenesis.

Sonic hedgehog (SHH), a protein involved in patterning of the early embryo, increases cardiogenesis when overexpressed in P19 cells (Gianakopoulos and Skerjanc 2005). The SHH effect is mediated through the expression of Gli2, which in turn activates the expression of BMP-4, a pro-cardiogenic factor, and of cardiac transcription factors MEF2C, GATA4, and Nkx2.5.

Hidai et al. (2003) studied the effect of FGF-1 and FGF-2 and showed that treatment of differentiating cells with FGF-1 induced the expression of cardiac markers (BMP-4, GATA-4, and Tbx5), as well as an endothelial marker (PECAM), whereas FGF-2 increased MyoD and noggin (a BMP antagonist) and abolished GATA-4 expression. These results suggest that FGF-1 has a positive effect on cardiogenesis by promoting lateral mesoderm specification. Figure 3A summarizes the different growth factors involved in cardiac differentiation in the mECC model.

Other types of factors can influence cardiac differentiation of P19 cells. Dynorphin B, a κ-opioid receptor agonist, induced Nkx2.5, GATA-4, and MHC expression (Ventura and Maioli 2000), while treatment of DMSO-primed P19 cells with a κ-opioid receptor antagonist abrogated cardiac differentiation. Oxytocin, an hypothalamic hormone involved in smooth muscle cell contraction during uterine contractions and lactation, induced cardiogenesis in the absence of DMSO and even allowed faster differentiation (cells beat at day 8 instead of day 12 in the presence of DMSO) (Paquin et al. 2002). The thyroid hormone triiodothyronine (T3) was also able to replace DMSO in inducing cardiac differentiation (Rodriguez et al. 1994).

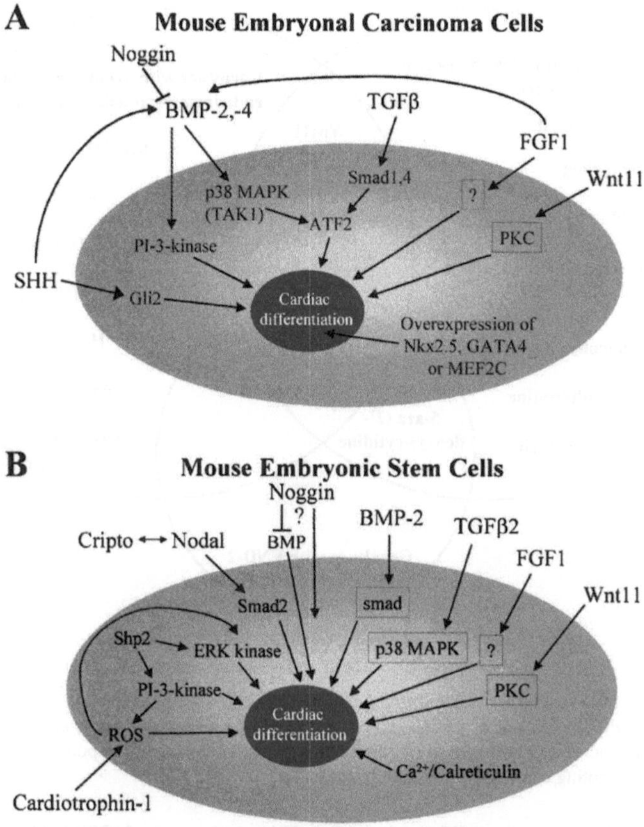

Fig. 3A, B Scheme of the growth factors playing a role in cardiac differentiation of mECCs (**A**) and mESCs (**B**) and their downstream cascades. All results were specifically obtained on mECC and mESC models, except for the pathways in frames (corresponding to the consensual signaling cascade)

Chemicals other than DMSO were tested for their ability to promote cardiogenesis of mECCs. The nucleoside analog, 5-azacytidine, can insert into DNA during replication, leading to hypomethylation and re-expression of genes. The effect of this drug was first described on NIH/3T3 cells, in which it induced their differentiation toward chondrocyte-, adipocyte- and skeletal myoblast-like phenotypes (Taylor and Jones 1979). Treatment with 5-azacytidine did not increase cardiogenesis in EBs, but, interestingly, it induced the expression of α-actinin and cardiac sarcomeric proteins, such as cardiac troponin T, in about 6% of cells in monolayer (Choi et al. 2004).

Finally, overexpression of cardiac transcription factors has been used to enhance cardiac differentiation of mECCs. Cells overexpressing Nkx2.5, MEF2C (Skerjanc et al. 1998), or GATA-4 (Grepin et al. 1997) differentiated toward beating cardiomyocytes without DMSO, indicating that each of these transcription factors can trigger the cardiac differentiation program. Furthermore, Monzen et al. (2002) confirmed that cells overexpressing wild-type Nkx2.5 could undergo DMSO-independent cardiogenesis, whereas cells overexpressing two mutant Nkx2.5 isoforms isolated from patients with cardiac malformations could not. On the other hand, cells overexpressing the T-box transcription factor TBX5 started to beat earlier,

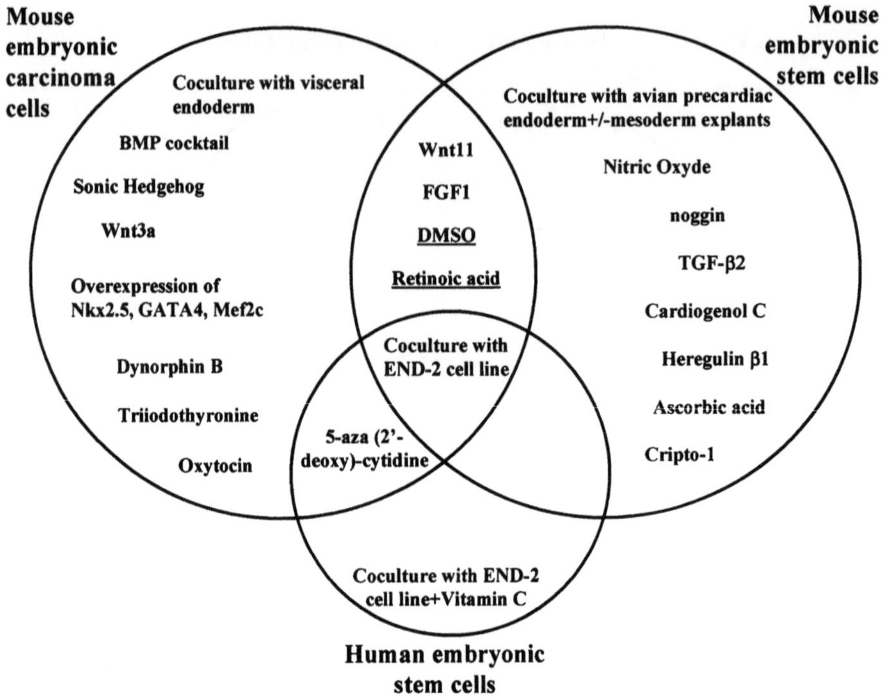

Fig. 4 Factors reported to enhance cardiac differentiation in ESC models. Some factors or compounds enhance cardiogenesis of mECCs, mESCs, or hESCs only, some enhance it in two cell types and only coculture with the END2 cell line increases cardiogenesis in the three models. *Underlined* compounds have been reported as ineffective in promoting cardiac differentiation of hESCs

but the amount of beating foci did not increase (Fijnvandraat et al. 2003; Hiroi et al. 2001). Factors or chemicals promoting cardiogenesis of mECCs are listed in Fig. 4.

Mouse embryonic stem cells

Mouse embryonic stem cell (mESC) lines were derived in 1981 from mouse blastocysts (Evans and Kaufman 1981). Their culture requires more care than mECCs and most of the cell lines have to be cultivated in the presence of feeder cells such as inactivated mouse embryonic fibroblasts (MEFs), to avoid spontaneous differentiation. In 1985, Doetschman et al. (1985) first reported the presence of cardiac cells in mESC-derived EBs.

Characterization of mESC-derived cardiomyocytes

Mouse ESC-derived cardiomyocytes have been characterized for their expression of different sarcomeric proteins over time. Guan et al. (1999) showed that the successive incorporation of several proteins into sarcomeres recapitulates myofibrillogenesis of the avian embryonic heart. Other studies indicate that β-MHC expression (usually observed in vivo during fetal life) starts at day 3–4, whereas α-MHC (usually present in vivo during adult life) is detected only from day 8 (Robbins et al. 1990; Sanchez et al. 1991). The analysis of tropomyosin isoform expression showed that striated muscle-specific α-tropomyosin is induced in differentiating EBs after 6 days. Striated muscle-specific β-tropomyosin was already present in undifferentiated cells, like several other non-muscle tropomyosin isoforms (Muthuchamy et al. 1993). Miller-Hance et al. (1993) studied the time-course expression of mlc and showed that mlc2v, the ventricular isoform, is expressed in EBs, suggesting that cells can mature in vitro to a ventricular-like phenotype. Concerning the expression of the different actin isoforms, smooth muscle, cardiac, and skeletal isoforms are expressed in mESC-derived cardiomyocytes and incorporated into sarcomeres (Ng et al. 1997). To study the effect of extracellular matrix on ultrastructural maturation of cardiomyocytes isolated from mEBs, cardiac cells were plated on plastic, matrigel, or cardiogel (a matrix deposited by cultured fibroblast isolated from neonatal mouse hearts) (Baharvand et al. 2005). Cells displayed a better sarcomeric organization on cardiogel, highlighting the fact that an extracellular matrix resembling in vivo conditions allows better in vitro maturation of the cells.

Electrophysiological studies revealed that mESC-derived cardiomyocytes display APs of embryonic and sinus node-like shapes and respond to chronotropic agents and channel blockers (Wobus et al. 1991). Moreover, single cell analysis by patch-clamp showed that APs mature over time. In early cardiomyocytes, only nodal-like APs are present, while later on, nodal-like, atrial-like, or ventricular-like APs can be recorded (Maltsev et al. 1993). Several other studies investigated specific electrophysiological aspects of mESC-derived cardiomyocytes, such as the appearance of ionic currents over time (Doevendans et al. 2000; Maltsev et al. 1994), response to β-adrenergic stimulation (Ali et al. 2004), changes in contractile sensitivity to Ca^{2+} (Metzger et al. 1994), T-type Ca^{2+} current characterization (Manabe et al. 2004; YM Zhang et al. 2003) and coupling of the Na^+/Ca^+ exchanger with Na^+/K^+ ATPase (Otsu et al. 2005). Characterization of cardiomyocyte subtypes using reporter genes under the control of different cardiac-specific promoters allowed enrichment of pacemaker and atrial cells using the α-MHC promoter (Kolossov et al. 2005) and of pacemaker cells using the atrial natriuretic peptide (ANP) promoter (Gassanov et al. 2004). In the latter study, treatment with endothelin-1 increased the percentage of pacemaker cells. As for ventricular cardiac cells, they can be isolated with a reporter gene under the control of the mlc2v promoter (Meyer et al. 2000).

Altogether, in vitro characterization of mESC-derived cardiomyocytes shows a clear correlation with the in vivo development of cardiac cells within the mouse embryo, validating the ESC differentiation system as an appropriate model for the study of developmental processes in vitro.

The potential of mESC-derived cardiomyocytes to repair a damaged myocardium has been assessed in vivo by transplantation studies. Klug et al. (1996) first injected selected mESC-derived cardiomyocytes in healthy myocardium of adult mdx mice (a strain lacking the dystrophin gene), underscoring the fact that injected dystrophin-positive cardiomyocytes could engraft and align their sarcomeres with host cells without teratoma formation. In a mouse model of myocardial infarction, transplantation of microdissected cardiomyocytes positive for green fluorescent protein (GFP) significantly improved heart function compared to control animals (Yang et al. 2002). However, purity of the injected cells and risk of teratoma formation are issues of crucial importance, as only two undifferentiated mESCs are sufficient to form a teratoma when injected subcutaneously in nude mice (Lawrenz et al. 2004).

Mouse ESCs: factors and pathways involved in cardiac differentiation

Many factors and morphogens known to play a role in cardiac differentiation in vivo (Fig. 1) have been tested on mESCs differentiating within EBs, either to confirm their involvement in the cardiogenic process or to increase the amount of cardiomyocytes. Factors necessary for or enhancing cardiac differentiation of mESCs are summarized in Figs. 3B and 4, respectively. Presence of endodermal cells within EBs is necessary for cardiogenesis to occur. In fact, Bader et al. (2001) showed that enzymatic removal of a primitive endodermal layer present around EBs after 2 days of differentiation in suspension prevents cardiac differentiation. In another study, coculture of early EBs with avian precardiac endoderm explants induced the presence of beating outgrowths in 65% of EBs, whereas only 10% of EBs cultured in control conditions (i.e., cultivated alone, on MEFs, or on nonprecardiac endoderm explants) showed spontaneous contractions (Rudy-Reil and Lough 2004). Interestingly, explants of endoderm and precardiac mesoderm together were even more effective as they induced 100% beating EBs, highlighting the importance of mesoderm-secreted factors. By coculturing mESCs with the END-2 cell line, results obtained on mECCs were confirmed on mESCs (Mummery et al. 2002).

Studies on specific endoderm-secreted factors have shown that treatment of mESCs with BMP-2 and TGF-β before EB formation induces the expression of Brachyury, Nkx2.5, and MEF2C mRNAs (Behfar et al. 2002). Furthermore, EBs derived from such pretreated cells contain increased beating areas, whereas beating is inhibited upon treatment with the BMP inhibitor noggin. The effect of noggin is, however, controversial. On one hand, continuous overexpression of noggin prevents cardiogenesis in mECC- (Monzen et al. 1999) and mESC-derived EBs (Behfar et al. 2002). In the same study, medium conditioned from a noggin expressing cell line inhibited cardiac differentiation of mESCs in coculture with neonatal cardiomyocytes. On the other hand, Yuasa et al. (2005) exploited the interesting observation that noggin is expressed in vivo during a short period of time in the mouse cardiogenic area. By pretreating mESCs with noggin before EB formation, the effect became pro-cardiogenic and the amount of beating EBs increased from 10% to more than 95%. Therefore, inhibition of BMP signaling by noggin might have an opposite effect on cardiogenesis, depending on when and how long the treatment is applied. Concerning the role of TGF-β, it was demonstrated that TGF-β2, but not -β1 or -β3, can increase the percentage of beating EBs (Kumar and Sun 2005).

Studies on FGF1-receptor knockout mESCs indicated that absence of the receptor reduced the percentage of beating EBs from 90% to 10%, demonstrating the importance of FGF signaling in cardiogenesis (Dell'Era et al. 2003).

Cripto is another growth factor expressed in the developing heart, which act as a cofactor in the signaling of a TGF-β family member named Nodal. Xu et al. (1998) demonstrated Cripto's role in cardiac differentiation, as its deletion in mESCs inhibited the appearance of contracting cardiomyocytes. The effect of Cripto on cardiogenesis is mediated through binding to the receptor Alk4 and activation of the Smad2 pathway (Parisi et al. 2003).

Involvement of the noncanonical Wnt pathway in cardiogenesis was demonstrated using a transgenic mESC line expressing GFP under the control of the Nkx2.5 promoter. Terami et al. (2004) reported that Wnt11-conditioned medium favored cardiogenesis, as the amount of GFP-positive cardiac cells within EBs increased from 7% to 15%.

Other studies have implicated pathways such as the heregulin-ErbB receptors pathway (Suk Kim et al. 2003) and the ephrin pathway (using ephrinB4 receptor-deficient mESCs) (Wang et al. 2004) in cardiomyocyte differentiation of mESCs.

Since, in most of these studies, differentiation experiments were conducted in the presence of 20% fetal calf serum (FCS), interference of growth factors present in animal serum might have biased results obtained by addition of exogenous growth factors. Sachinidis et al. (2003) developed a serum-free differentiation protocol by switching at day 5 of EB culture from 20% FCS to a serum replacement medium containing BSA, insulin, and transferrin, but no growth factors. Under these conditions, 80% of EBs contained beating areas and this proportion rose to 100% with addition of PDGF-BB. However, the obtained cardiac cells did not display a high degree of sarcomeric organization, indicating their still immature phenotype.

The mESC model also confirmed the involvement of other signaling pathways in cardiac differentiation. Studies on Shp2-deficient mESCs showed that absence of this protein tyrosine phosphatase, ubiquitously expressed during mouse development and implicated in regulation of signaling events downstream of growth factor receptors, decreased and delayed cardiac differentiation within EBs (Qu and Feng 1998). Shp2 can activate the ERK kinase pathway as well as the PI3-kinase pathway. Treatment of EBs with the PI-3-kinase inhibitor LY294002 slowed down EB growth and markedly decreased beating areas (Klinz et al. 1999). This effect could be due to a direct involvement of PI-3-kinase signaling on proliferation and/or survival of cardiac precursor cells, rather than on the differentiation process.

A possible consequence of PI-3-kinase activation is the generation of reactive oxygen species (ROS). A study by Sauer et al. (2000) reported that 2- to 3-day-old EBs generate measurable amounts of ROS and that inhibition of endogenous ROS production by free radical scavengers decreases the percentage of contracting areas. A similar effect was obtained using PI-3-kinase inhibitors. This phenotype was rescued by adding exogenous ROS to PI-3-kinase inhibitor-treated EBs. The same group showed that cardiotrophin-1 is one of the endogenous factors promoting survival and proliferation of cardiac cells within EBs through generation of ROS (Sauer et al. 2004). Indeed, treatment of differentiating EBs with cardiotrophin-1 increased ROS levels and activated a signaling cascade involving JAK2, NF-κB, STAT3, and ERK. Other studies have confirmed the role of JAK2 and STAT3 in mESC-derived cardiogenesis (Foshay et al. 2005).

Other intracellular mechanisms, such as Ca^{2+}-induced signaling, have been implicated in cardiac differentiation. The endoplasmic reticulum chaperone calreticulin (CRT) plays a role in cardiogenesis, as cardiac differentiation is impaired in EBs derived from CRT-deficient mESCs, mainly because of disorganized myofibrillogenesis (Li et al. 2002). This phenomenon was rescued by a transient ionomycin-induced elevation of the intracellular free Ca^{2+} concentration in CRT-deficient EBs. Similarly, inhibition of Ca^{2+}/calmodulin-dependant kinases (CAMKs) in wild-type EBs mimicked the phenotype of CRT-deficient cells. This indicates that a Ca^{2+} checkpoint is essential to activate CAMKs that, in turn, induce nuclear translocation of MEF2C and mlc2v expression, phosphorylation, and incorporation into functional sarcomeres (for review see Puceat and Jaconi 2005).

Small molecules known to enhance cardiac differentiation in mECCs were tested on mESCs. DMSO induced myocytic differentiation, but was not specific for cardiogenesis, as demonstrated by the presence of skeletal and smooth muscle cells together with cardiac cells (Dinsmore et al. 1996). Unlike mECCs, the effect of retinoic acid (RA) on mESC differentiation toward cardiac cells is controversial and appears highly time- and concentration-dependent. Concentrations higher than 10^{-9} M seem to decrease expression of mesodermal and cardiac genes and favor neuronal differentiation (Bain et al. 1996; Dinsmore et al. 1996). However, other studies report either that 10^{-9} M RA does not modify the number of beating EBs (Dani et al. 1997) or that it can increase this number (Wobus et al. 1997). In the latter study, a switch in cardiac phenotypes was also observed as the amount of cardiomyocytes with Purkinje and ventricular phenotypes was increased compared to atrial and nodal cells. Interestingly, treatment of EBs with 10^{-6} M RA activates the ERK signaling pathway, which is a candidate pathway implicated in cardiac differentiation, but these EBs do not beat (Bost et al. 2002). Treatment of EBs with PD98059, an inhibitor of MEK1 (the ERK-activating kinase), had no effect on the beating percentage either. Therefore, the RA effect on cardiac differentiation seems to be mediated through activation of an ERK-independent pathway.

It has been documented that CD44, the hyaluronan receptor, is expressed in the embryonic heart (Wheatley et al. 1993) and that upon binding, hyaluronan is internalized and can act as a carrier. Treatment of EBs with hyaluronan conjugated to RA and butyric acid induced a significant increase in the beating clusters, whereas hyaluronan conjugated to only one of the compounds was not as effective and hyaluronan alone decreased the number of beating foci (Ventura et al. 2004).

Other approaches have been used to discover pro-cardiogenic molecules. For example, Takahashi et al. (2003) investigated the effect of 880 chemicals approved for human use on mESCs cultured as monolayers instead of EBs. Using mESCs stably transfected with GFP under the control of the α-MHC promoter, they found that ascorbic acid (vitamin C) induced five times more contractile GFP-positive cells compared to control conditions, whereas other anti-oxidants such as N-acetylcysteine or vitamin E had no effect. In another study, a library of heterocycles was screened for the ability to induce cardiac differentiation of mESCs in monolayers (Wu et al. 2004). One compound in particular, named cardiogenol C, led to differentiation into beating myocytes after 7 days of culture and 40–55% of the cells stained positive for MHC.

Nitric oxide (NO) is a small molecule produced by several isoforms of the nitric oxide synthase (NOS) and can regulate cellular processes through guanylate cyclase activation and cGMP production. Its involvement in cardiac differentiation processes was reported as several isoforms are expressed in early embryonic heart and in differentiating EBs (Bloch et al. 1999). In addition, NOS inhibitors decreased the myofibrillar organization of mESC-derived cardiomyocytes, without changing their number. Another study demonstrated that addition of exogenous NO or overexpression of NOS increased the percentage and the surface of contracting foci as well as the expression of cardiac-specific proteins (Kanno et al. 2004).

Human embryonic stem cells

In 1998, Thomson et al. (1998) derived the first human embryonic stem cell (hESC) lines from in vitro-fertilized supernumerary embryos. Because fetuses from abortions were previously the only source of human prenatal cardiomyocytes, hESCs are a unique tool to study human cardiogenesis in vitro. Their derivation raised great hope and hype for cell replacement therapies and heart regeneration. Since then, more than a hundred cell lines have been derived worldwide (Hoffman and Carpenter 2005). Classically, hESC lines are cultured over mitotically inactivated MEFs in order to maintain them in an undifferentiated state. The presence of feeder cells might also be important for other reasons such as avoiding selection pressure that can lead to karyotypic changes (Draper et al. 2004).

Induction of differentiation has been first triggered using the same technique than for mECCs and mESCs, i.e., EB formation. Because hESCs grow as compact colonies and do not tolerate isolation into single cells, EBs are formed by placing colonies in suspension in the presence of FCS, leading to random aggregation. This technique allows the cells to spontaneously differentiate. The first demonstration that hESCs could differentiate toward cells of the three germ layers, including beating cardiomyocytes, was reported in 2000 (Itskovitz-Eldor et al. 2000).

Characterization of hESC-derived cardiomyocytes

The first characterization of hESC-derived cardiomyocytes showed that only 8% of all EBs contained beating areas after 30 days of differentiation. They expressed cardiac genes (such as Nkx2.5, GATA4, cardiac troponins T and I, ANP, α-MHC, mlc2v, and mlc2a) at the mRNA level (Kehat et al. 2001). By immunostaining, sarcomeric structures were observed with anti-cardiac troponin I and anti-MHC antibodies. The cells also displayed Ca^{2+} transients and their beating rate was sensitive to chronotropic agents. In another study, the presence of cardiomyocytes was observed in up to 68% of EBs after 15 days of differentiation (Xu et al. 2002). Differences might be explained by the use of different cell lines, cell propagation conditions, or protocols to induce differentiation.

Ultrastructural maturation of the cells was then assessed by electron microscopy. The cardiomyocytes displayed an increasing content of sarcomeric proteins, whose organization became more and more complex over time (Snir et al. 2003). Their size also increased over time and they divided until day 35 of differentiation. However, no T-tubules were observed even after 60 days, indicating that the cells do not develop to an adult phenotype.

Several functional studies have already implemented the characterization of hESC-derived cardiomyocytes. He et al. (2003) recorded nodal- and both embryonic atrial- and ventricular-like APs in beating outgrowths. Additionally, one major AP phenotype was generally found within the same outgrowth, suggesting that a predominant cell type might differentiate within an outgrowth.

Moreover, functionality of the muscarinic and β-adrenergic signaling pathways was demonstrated in human cardiac cells (Reppel et al. 2004). Na^+ and pacemaker currents were recorded in these cells, but not inward rectifier K^+ current, suggesting that Na^+ current is an important player in triggering APs (Satin et al. 2004).

Monitoring of the conduction properties in EB outgrowths plated on a multi electrode array revealed the presence of areas with different conduction velocities (Kehat et al. 2002). A fast and a slow group were observed, depending on the 3D micro-architecture of each EB, as presence of narrow strands of conducting tissue can slow down the influx. The velocity values measured in these experiments were slower than in human heart, or rat and mouse cardiomyocyte monolayers.

The ability of hESC-derived cardiomyocytes to couple electromechanically with other cardiac cells was studied in a coculture model with rat neonatal ventricular cardiomyocytes (Kehat et al. 2004). The rat cells initiated electrical impulses that propagated to the human myocytes. In the same study, in vivo integration of the cells was demonstrated in a swine model of atrioventricular block. Indeed, implantation of outgrowths containing human cardiomyocytes could restore an escape rhythm.

In a model of coculture with quiescent rat neonatal cardiomyocytes, Xue et al. (2005) showed that, by direct contact, dissected human cardiomyocytes restored the electrical impulse and paced the rat cells. In vivo, outgrowths were able to pace guinea pig hearts that had undergone cryoablation of the atrioventricular node, confirming results from the previous study. These data indicate the feasibility, at least at short time points, of using hESC-derived cardiomyocytes as cell-based pacemakers.

Human ESCs: factors and pathways involved in cardiac differentiation

Ongoing studies using the hESC model will confirm or infirm the role of growth factors and morphogens, unraveled with other models (Figs. 1 and 3), that are also involved in human cardiac differentiation. Strategies to enhance cardiogenesis in hEBs are currently under intense investigation. The effect of several growth factors on cells dissociated from 5 days old EBs was assessed (Schuldiner et al. 2000). Activin A and TGFβ1 turned on the expression of mesoderm genes only, whereas retinoic acid, BMP-4, bFGF, and EGF turned on mesodermal and ectodermal markers and NGF and HGF induced markers from the three embryonic layers.

As seen with mECCs or mESCs, a coculture system with the visceral endoderm-like cell line END2 induced cardiac differentiation without the need of EB formation (Mummery et al. 2003). The cells obtained with this protocol also displayed nodal-, atrial-, and ventricular-like APs. Surprisingly, using the same coculture model, removal of FBS induced a 24-fold increase in the number of beating areas (Passier et al. 2005). Therefore, endodermal factors secreted by the END2 cell line are sufficient to induce cardiogenesis, while FBS, traditionally used in EB differentiation experiments, might contain cardiac inhibitory factors.

Molecules known to induce cardiogenesis in mECCs and mESCs were tested on differentiating hEBs. Neither DMSO nor RA increased the percentage of beating EBs (Kehat et al. 2001). However, treatment of EBs from day 6 to 8 of differentiation with 5-aza-2'-deoxycytidine increased the amount of α-MHC (Xu et al. 2002). Interestingly, addition of ascorbic acid in the END2-hESC coculture system, in the absence of FBS, further increased the number of beating areas by 40% (Passier et al. 2005). Figure 4 summarizes factors or chemical compounds that can increase cardiac differentiation of hESCs.

Lessons from embryo-derived stem cells

Studies on mECCs, mESCs, and hESCs bring new insights on cardiac development that might lead to a future use of hESCs in transplantation therapies. Figure 3 illustrates growth factors and their corresponding signaling cascades playing a role in cardiac differentiation of mECCs and mESCs. Mouse ECCs have for the first time allowed the study of cardiac differentiation in vitro and a better characterization of transcription factors interactions, gene expression, cell–cell interaction and growth factors required for cardiogenesis (Fig. 3A). Differentiation of mESCs into cardiac myocytes allowed the accurate characterization of growth factors effects and functional maturation of cardiomyocytes during development (Fig. 3B; for other reviews see Boheler et al. 2002, 2005; Foley and Mercola 2004). Indeed, some factors might have opposite effects depending on the time of treatment and the concentration (for example, BMP-2/noggin or RA). Studies on hESCs are still at an early stage. Whether growth factors and signaling cascades involved are similar or distinct between the mouse and human ESC models is not known yet. Upscaling the production of hESC-derived cardiomyocytes and assessing their behavior and function in animal models are the next fundamental steps in view of future therapeutic use of these cells.

Variability in results with ESC studies can be explained by several nonexclusive facts. First, the percentage of beating EBs in control conditions can be very different depending on cell lines, protocols, culture medium, or FBS composition. This could explain why a factor enhances cardiogenesis only in specific conditions. Secondly, differentiating EBs contain a mixed cell population, unless the cells of interest are specifically purified. Therefore, depending on the technique used, there is no certitude that a studied gene is specifically expressed or upregulated in the cardiac cell population.

Working with purified mesodermal progenitor cells could undeniably circumvent these technical concerns and allow testing the effect of growth factors on purified cell populations. For example, early mesodermal progenitor cells were isolated using mESCs containing GFP knocked in the brachyury gene, an early mesodermal transcription factor (Kouskoff et al. 2005). These cells were able to differentiate toward hematopoietic and cardiac cells. Isolation and amplification of cardiac progenitor cells from differentiating EBs have been also recently reported using selection with the mesodermal marker Flk1 followed by a coculture step with OP9 stromal cells (Iida et al. 2005).

Both mouse and human ESC models closely recapitulate in vivo cardiac development, at a structural and electrophysiological level. However, sarcomeric organization does not reach an adult phenotype in any of the models. In both models, electrophysiological recordings showed a maturation of the cells over time. Major differences between mouse and human cells, highlighted by several groups, are the lower occurrence of beating areas in human EBs, the lower beating rate of human cardiomyocytes and the slower differentiation, most probably fitting the human developmental timeframe.

Strategies to promote cardiac differentiation by treating EBs or monolayers of undifferentiated cells with pro-cardiogenic factors give promising results to concoct in the future a "cardiogenic cocktail" that could induce differentiation into a pure population of cardiac cells. Figure 4 summarizes what enhances cardiac differentiation of mECCs, mESCs, and/or hESCs. All the factors and compounds used can promote cardiogenesis at different levels, by promoting endoderm and mesoderm formation, cardiac precursor cell proliferation and/or survival. Most of the factors shown to increase cardiac differentiation have been reported for one cell type only. It is interesting to note that only the coculture with the END-2 cell line seems to increase cardiogenesis in the three cell types, highlighting the fact that a combination of several growth factors, activating and/or inhibiting, is probably necessary.

Adult stem cells

The adult body contains stem cell populations whose role is to participate in the turnover and healing of organs. For example, hematopoietic stem cells produce all of the blood cells, epithelial stem cells regenerate epithelia, and satellite cells participate in healing of skeletal muscles. Many studies have demonstrated that these adult stem cells have a differentiation potential restricted to their organ of origin, i.e., they are unipotent or multipotent. However, recent reports shake dogmas about limitations of their differentiation potential (for review see Lakshmipathy and Verfaillie 2005; Quesenberry et al. 2005), giving hope to using cells from patients themselves for cell therapy. Indeed, the concept of transdifferentiation, or lineage conversion, involves the ability of an adult stem cell to become a cell with an unexpected phenotype (Wagers and Weissman 2004).

Many studies assessed the in vivo potential of adult stem cell transplantation to rescue the function of damaged hearts, testing the hypothesis that local environment could drive appropriate differentiation. Although animal studies give promising results, the mechanism that allows improvement of the heart function is not clear yet. Indeed, cell fusion phenomenon, elastic tension improvement of the ventricular wall, or neoangiogenesis can occur and would not involve a direct cardiac differentiation of the transplanted cells. Adult stem cells have therefore been tested for their ability to differentiate toward a cardiac phenotype in vitro.

Bone marrow-derived stem cells

Several types of adult stem cells reside in the bone marrow (Fig. 2). First, hematopoietic stem cells (HSCs) give rise to all lineages of the blood. Grafts of bone marrow containing HSCs or specifically isolated HSCs have been used for treatment of leukemia for several decades and were the first cell therapy procedure used in clinical settings. Second, mesenchymal stem cells (MSCs) (Friedenstein et al. 1976), also known as bone marrow stromal cells or colony-forming unit fibroblasts, are the in vivo precursors of osteoblasts, chondrocytes, and adipocytes (Pereira et al. 1995). Third, the group of Verfaillie (Reyes and Verfaillie 2001) isolated a subpopulation of MSCs from the bone marrow, named multipotent adult progenitor cells (MAPCs), which seem able to differentiate toward cells of the three germ layers. The role of these cells in vivo is not clear so far.

Hematopoietic stem cells

HSCs' innate differentiation potential comprises all cellular lineages of the blood. Their potential to differentiate into cardiomyocytes has been thoroughly investigated in vivo. Until now, results have been highly controversial. Several studies indicated that bone marrow cells enriched in HSCs can regenerate myocardium in animal models of myocardial infarction (Agbulut et al. 2003; Jackson et al. 2001; Kajstura et al. 2005; Orlic et al. 2001a,. 2001b). However, recent reports showed that the same cell populations transplanted in the myocardium adopted only an hematopoietic phenotype (Balsam et al. 2004; Murry et al. 2004). Other strategies such as mobilization of hematopoietic precursor cells by stem cell factor and granulocyte-colony stimulating factor during the course of myocardial infarction gave contradictory results as well (Deten et al. 2005; Orlic et al. 2001c).

To date, there is therefore no consensus regarding the differentiation potential of hematopoietic stem cells toward cardiomyocytes in vivo. Many factors can explain these contradictory results. First, the purification of the injected cells varies among studies. HSCs were enriched by depletion of all mature hematopoietic cells and selection of the remaining cells with stem cell surface markers Sca1 and/or c-kit. On the other hand, other groups used unpurified bone marrow, as it is not clear yet which cells have the potential to become cardiomyocytes. Second, variable cell numbers are injected using different administration routes such as intramyocardial, intracoronary, or intravenous. Third, hearts are analyzed at different time points after injection. Fourth, some models are xenogenic or allogenic, which can increase or decrease immunorejection.

An important issue is to determine whether donor-derived cells are found in the myocardium because of fusion events with endogenous cardiomyocytes or because of their transdifferentiation into cardiac cells. In some studies, the phenomenon only represented cell fusions (Alvarez-Dolado et al. 2003; Kajstura et al. 2005; Lapidos et al. 2004; Nygren et al. 2004; Terada et al. 2002). In other reports, both fusion and transdifferentiation occurred (Zhang et al. 2004) or transdifferentiation only (Kajstura et al. 2005).

Little evidence exists concerning the in vitro potential of hematopoietic progenitor cells to generate cardiomyocytes. When cultured in the presence of dexamethasone, RA, prostaglandin E2, interleukin 2, and BMP-4, hematopoietic progenitor cells from chick bone marrow expressed several cardiac transcription factors and sarcomeric proteins (Eisenberg et al. 2003). However, the cells did not contract spontaneously.

Mesenchymal stem cells

MSCs located in the bone marrow are classically isolated by their adherence to plastic (Friedenstein et al. 1976). They can differentiate toward osteoblasts, chondrocytes, and adipocytes in vivo (Pereira et al. 1995) as well as in vitro (Pittenger et al. 1999). Successful isolation of MSCs or MSC-like cells from other organs includes fat tissue (Zuk et al. 2001), umbilical cord (Romanov et al. 2003), and amniotic fluid, placenta, or amnion (In 't Anker et al. 2004).

Like cardiomyocytes, MSCs derive from mesoderm and it is therefore of great interest to investigate their potential to be directed toward a cardiac phenotype.

In vitro studies have shown that 5-azacytidine treatment induced human MSCs to differentiate into nonbeating cardiomyocytes (Xu et al. 2004). With mouse MSCs, Makino et al. (1999) observed spontaneously beating cardiomyogenic cells 3–4 weeks after treatment, though at a variable frequency. These cells expressed cardiac sarcomeric proteins and functional adrenergic and muscarinic receptors and displayed sinus node-like and ventricular-like APs (Hakuno et al. 2002). However, other groups could not reproduce these results (Cao et al. 2004; Liu et al. 2003) or only by coculturing MSCs with neonatal cardiomyocytes (Rastan et al. 2005; Takeda et al. 2004). Nevertheless, when a coculture step is added, cell fusion events have to be undoubtedly ruled out (Camargo et al. 2004), as it was shown that isolated neonatal cardiomyocytes can fuse with cardiac fibroblasts and endothelial cells, even though at a very low frequency (Matsuura et al. 2004).

Other strategies include overexpression of myocardin, a transcription factor involved in cardiac and smooth muscle differentiation, which could activate in human MSCs the expression of cardiac and smooth muscle genes (van Tuyn et al. 2005). HCN2, an ion channel expressed in pacemaker cells, was also overexpressed in MSCs which then displayed pacemaker-like currents (Potapova et al. 2004). Upon transplant in dog left ventricle, these cells formed gap junctions with the surrounding cardiomyocytes and induced a ventricular spontaneous escape rhythm when the sinusal rhythm was blocked.

In summary, few demonstrations exist concerning the potential of MSCs to differentiate into cardiac cells in vitro and the efficiency of the process is low.

Transplantation of MSCs in the heart of several animal models seems to ameliorate heart function, but whether it is by the de novo formation of cardiomyocytes or by a pro-angiogenic effect remains to be demonstrated (Gojo et al. 2003; Nagaya et al. 2004; Toma et al. 2002). As cells survive poorly to direct intramyocardial injection, rat MSCs were transduced with Akt1, a serine threonine kinase known to have a pronounced pro-survival effect. Once injected in a rat model of myocardial infarction, such cells regenerated 80–90% of lost cardiac volume (fourfold better than wild-type MSCs) (Mangi et al. 2003). Furthermore, the same group showed recently that this phenomenon might be due to a paracrine effect, as medium conditioned from MSCs overexpressing Akt1 gives the same results (Gnecchi et al. 2005). Nevertheless, trials have been initiated in patients with recent myocardial infarction. Autologous bone marrow cells were injected into the left coronary artery and a 6-month follow-up shows an improvement in heart function compared to control patients (Chen et al. 2004; Wollert et al. 2004).

As MSCs are a heterogeneous cell population, purification of homogenous MSC populations are attempted using cell surface markers. Stro-1 is one of the candidate markers (Simmons and Torok-Storb 1991), since Stro-1-positive human cells are able to differentiate toward adipocytes, chondrocytes, osteoblasts, and vascular smooth muscle-like phenotypes (Dennis et al. 2002). However, anti-Stro-1 antibodies also recognize erythroid precursor cells. Three other antibodies—SH2, SH3, and SH4—were produced against human MSCs

(Haynesworth et al. 1992). SH2 reacts with endoglin (Barry et al. 1999), a glycoprotein associating with TGFβ receptors, whereas SH3 and SH4 recognize CD73 (Barry et al. 2001), a GPI-anchored glycoprotein usually expressed on lymphoid cells. However, both endoglin and CD73 are not exclusively present on MSCs. To date, despite the possibility of enriching the MSC fraction through different surface antibodies, no marker is available to positively isolate pure MSCs.

Multipotent progenitor cells

MAPCs, an interesting cell population possibly able to differentiate toward cells of the three germ layers, have been isolated from bone marrow (Reyes and Verfaillie 2001). The isolation protocol includes coating of culture dishes with fibronectin and use of a medium containing 2% FBS, EGF, PDGF-BB, and LIF. Two or 3 weeks after isolation, a depletion step with anti-CD45 and anti-glycophorin A antibodies allows the removal of hematopoietic and red blood cells, respectively. Whereas primary cells usually proliferate in vitro for short periods of time, MAPCs can be cultured for more than 120 population doublings. They express Oct4 (a pluripotency marker expressed in ESCs) and upon injection into a mouse blastocyst, can generate up to 45% chimerism, including in the heart (Jiang et al. 2002). Since then, other groups have reported the isolation of multipotent cells from bone marrow (Yoon et al. 2005) or cord blood (Kogler et al. 2004). The ability of these multipotent progenitor cells to become cardiac cells has not yet been demonstrated in vitro, but as their differentiation potential seems wider, more investigations are needed.

Skeletal muscle-derived stem cells

Skeletal muscle contains a well-characterized stem cell population called satellite cells. Their role in vivo is to regenerate damaged skeletal myotubes. As they are programmed to differentiate toward skeletal muscle, their potential to switch to a cardiac phenotype upon injection in vivo has been thoroughly investigated. Satellite cells were injected in several animal models of myocardial infarction and have been shown to ameliorate heart function (Al Attar et al. 2003; Blatt et al. 2003; Horackova et al. 2004; Scorsin et al. 2000). Trials in patients undergoing coronary bypasses have also started and injection of the own patient's satellite cells also seems to improve heart function (Menasche et al. 2001, 2003; Siminiak et al. 2004; Zhang et al. 2003). However, no evidence of muscle phenotype switch from skeletal to cardiac was demonstrated (Dorfman et al. 1998; Reinecke et al. 2002). Intracardiac myotubes do not appear to be connected with the surrounding myocardium (Leobon et al. 2003), except on very rare occasions (Rubart et al. 2004). Therefore, satellite cells might ameliorate heart function through other mechanisms such as passive improvement of the ventricular wall tension, limitation of scar expansion, or extracellular matrix remodeling.

Interestingly, Winitsky et al. (2005) isolated from skeletal muscle a stem cell population distinct from satellite cells. These cells are negative for stem cell markers Sca-1 and c-kit and can grow in suspension. They are able to express cardiac genes and differentiate in vitro toward beating cardiac cells exhibiting calcium transients and APs. As no differentiation of satellite cells toward cardiac myocytes could be shown in vivo, this new population of cells might represent an interesting alternative.

Cardiac stem cells

The mammalian heart had always been thought to be a postmitotic organ containing no stem cell population, mainly because of the lack of regeneration observed in vivo after damage. Indeed, when a myocardial infarction occurs, cardiomyocytes die and are replaced by an akinetic fibrotic tissue. However, some groups have claimed for years that cardiomyocytes can still divide in the adult heart, though at a very low rate (Beltrami et al. 2001; Reiss et al. 1994). Therefore, attempts have been made recently to isolate cardiac stem cells from adult hearts.

Several groups have reported the isolation of stem cell populations from postnatal hearts using techniques or surface markers previously developed to isolate stem cells from other organs. Some stem cells do express the multi-drug resistance (MDR) gene and are called side population (SP) cells, as they have the ability to exclude Hoechst dye, enabling their sorting by flow cytometry. In mice, SP cells isolated from the myocardium give rise to beating colonies in vitro when cocultured with rat primary cardiomyocytes (Hierlihy et al. 2002). Oh et al. (2003, 2004) showed that SP cells display a new surface marker phenotype (Sca-1$^+$, CD31$^+$, and CD38$^+$), which excludes them from being hematopoietic stem cells, endothelial progenitor cells, or mature endothelial cells. Telomerase activity is present in these cells, as well as expression of the cardiac transcription factors MEF2C, GATA4, and TEF-1. In vitro treatment with 5-azacytidine induces the expression of Nkx2.5 and sarcomeric proteins and, when injected in a mouse model of myocardial infarction, these cells home and start to differentiate in the infarct border zone. Furthermore, upon injection into a mouse blastocyst, a cardiac chimerism is observed in 33% of the animals (Oh et al. 2004). Recently, Sca-1$^+$ cardiac precursor cells were shown to proliferate in vitro and to differentiate into cardiomyocytes over a 3-week period, upon treatment with dexamethasone, β-glycerophosphate, and ascorbic acid (Rosenblatt-Velin et al. 2005). Five percent of the cells turned into spontaneously contracting troponin I-positive cells. Moreover, FGF2 seems to be necessary for this process, as Sca-1$^+$ cells isolated from mice lacking the FGF2 gene do not differentiate in the same conditions.

Another approach was developed to isolate potential cardiac stem cells from human and mouse hearts (Messina et al. 2004). Upon culture of cardiac explants, cells migrated out and formed aggregates named cardiospheres. These cells started to spontaneously contract in vitro and expanded exponentially for more than 50 days. Another stem cell marker, c-kit, was used to identify in vivo and purify cardiac stem cells from the rat heart (Beltrami et al. 2003). Clonogenic c-kit-positive cells could give rise to cardiac, smooth muscle, and endothelial cells in vitro as well as in vivo, when injected in the border zone of an infarcted area. The same c-kit-positive cells were found in vivo in human hearts and their number increased in cardiac hypertrophy conditions (Urbanek et al. 2003). The same authors recently reported that both early or end-stage infarcted hearts contain more cardiac stem cells than control hearts and that these c-kit$^+$ cells have shorter telomeres and are more often positive for apoptosis or senescence markers (Urbanek et al. 2005).

Recently, isl1, a transcription factor important for Langerhans islet differentiation, was also shown to be important for cardiac differentiation, as mice lacking this gene do not develop the outflow tract, right ventricle, and atria (Cai et al. 2003). Isl1 is expressed as well in some cells of the postnatal heart and was therefore used to isolate cardiac precursor cells from human, rat, and mice. Indeed, the adult heart contains an isl1-positive cell population able to differentiate in vitro toward mature cardiac cells at a frequency of 25% when cocultured with neonatal myocytes (Laugwitz et al. 2005).

As several research groups isolated cardiac precursors using different culture protocols and cell surface markers, it is not yet possible to make a claim for the existence of a single cardiac stem cell population. However, some features seem to be common among the different protocols, such as culture of the cells in low serum conditions and the need of fibroblast-like or mesenchymal-like feeder cells for primary cell propagation.

In summary, recent reports suggest the presence of resident cardiac stem cells in the adult heart. Although insufficient to repair the myocardium upon major injury, their recruitment and/or ex vivo expansion before reinjection might represent an alternative option.

Lessons from adult stem cells

Altogether, there are some reports of in vitro differentiation of adult stem cells toward cardiomyocytes. However, the yield of cardiac cells obtained is clearly much lower than with ESCs and few studies provide a functional electrophysiological characterization of the obtained myocytes. On the other hand, several groups observed cells that re-express cardiac markers and that may be on the way to cardiac differentiation. Future studies are needed to show whether one can push these cells to develop into fully differentiated functional cardiomyocytes using cardiogenic factors (Figs. 1 and 3).

To achieve this goal, a better understanding of fundamental adult stem cell biology is needed. Most stem cell populations currently studied are heterogenous and display different surface marker expression profiles. Generation of antibodies to specifically sort them is thus mandatory. Several groups are currently working at the clonal level to unambiguously prove selective differentiation toward unexpected phenotypes. Answering these questions is of crucial importance as some precursors might have a wider differentiation potential than others.

Conclusion

Embryo-derived stem cells provide a unique tool to study cardiogenesis in vitro. Much progress has been made in the understanding of cardiac differentiation mechanisms, especially in deciphering the role of growth factors. Mouse ECC-, mESC-, and hESC-derived cardiomyocytes have been characterized functionally and structurally and although their maturation state remains at an embryonic or neonatal stage, their development recapitulates early in vivo cardiogenesis. The amounts of cardiomyocytes obtained are sufficient for in vitro studies. Nevertheless, upscaling the production by treatment with growth factors is needed for heart cell therapy purposes. Better protocols are likewise needed for cell purification, as the potential of ESCs to form teratomas must not be neglected. Further studies should aim at better understanding how the cells (differentiated or not) respond to the environment upon implantation in small and large animal models, both from a functional and an immunogenic point of view.

The potential for adult stem cells to generate cardiomyocytes in vitro is still under investigation. There is no consensus yet on which adult stem cells have a clear potential to become cardiac cells. However, the knowledge gained with ESC studies will help directing adult stem cells in vitro toward the cardiac phenotype. Autologous adult stem cells are a source of cells to consider very seriously, as their use would circumvent many problems encountered by ESC-based therapies, such as immunocompatibility and teratoma formation.

Future studies on stem cells using large-scale genomic and proteomic analysis will help to unravel molecular determinants regulating differentiation processes. Other potential areas of research using stem cells include the study of fundamental processes such as reprogramming and epigenetic modifications. The effect of up- or downregulation of genes and mutated proteins responsible for human diseases can be studied in stem cell models. Pharmacological studies on ESCs will also be useful to test the safety and teratogenicity of chemicals.

Embryonic- and adult-stem cell research are complementary fields. At this point, no option must be overlooked, as using adult stem cell-derived cardiomyocytes for heart cell therapy would avoid immunorejection and ethical issues, two major concerns against the use of ESCs for such treatments. They should progress in parallel in order to eventually lead to the development of efficient cell transplantation treatments for heart failure.

Acknowledgements. We thank the Swiss Academy of Medical Sciences (No. 14/03, EB's salary) and the Swiss National Science Foundation (No. 3236B0–104480) for financial support. A special thanks is extended to Dr. Karen Bedard for critical reading of the manuscript.

References

Agbulut O, Menot ML, Li Z, Marotte F, Paulin D, Hagege AA, Chomienne C, Samuel JL, Menasche P (2003) Temporal patterns of bone marrow cell differentiation following transplantation in doxorubicin-induced cardiomyopathy. Cardiovasc Res 58:451–459

Al Attar N, Carrion C, Ghostine S, Garcin I, Vilquin JT, Hagege AA, Menasche P (2003) Long-term (1 year) functional and histological results of autologous skeletal muscle cells transplantation in rat. Cardiovasc Res 58:142–148

Ali NN, Xu X, Brito-Martins M, Poole-Wilson PA, Harding SE, Fuller SJ (2004) Beta-adrenoceptor subtype dependence of chronotropy in mouse embryonic stem cell-derived cardiomyocytes. Basic Res Cardiol 99:382–391

Alvarez-Dolado M, Pardal R, Garcia-Verdugo JM, Fike JR, Lee HO, Pfeffer K, Lois C, Morrison SJ, Alvarez-Buylla A (2003) Fusion of bone-marrow-derived cells with Purkinje neurons, cardiomyocytes and hepatocytes. Nature 425:968–973

Bader A, Gruss A, Hollrigl A, Al-Dubai H, Capetanaki Y, Weitzer G (2001) Paracrine promotion of cardiomyogenesis in embryoid bodies by LIF modulated endoderm. Differentiation 68:31–43

Baharvand H, Azarnia M, Parivar K, Ashtiani SK (2005) The effect of extracellular matrix on embryonic stem cell-derived cardiomyocytes. J Mol Cell Cardiol 38:495–503

Bain G, Ray WJ, Yao M, Gottlieb DI (1996) Retinoic acid promotes neural and represses mesodermal gene expression in mouse embryonic stem cells in culture. Biochem Biophys Res Commun 223:691–694

Balsam LB, Wagers AJ, Christensen JL, Kofidis T, Weissman IL, Robbins RC (2004) Haematopoietic stem cells adopt mature haematopoietic fates in ischaemic myocardium. Nature 428:668–673

Barry F, Boynton R, Murphy M, Haynesworth S, Zaia J (2001) The SH-3 and SH-4 antibodies recognize distinct epitopes on CD73 from human mesenchymal stem cells. Biochem Biophys Res Commun 289:519–524

Barry FP, Boynton RE, Haynesworth S, Murphy JM, Zaia J (1999) The monoclonal antibody SH-2, raised against human mesenchymal stem cells, recognizes an epitope on endoglin (CD105). Biochem Biophys Res Commun 265:134–139

Behfar A, Zingman LV, Hodgson DM, Rauzier JM, Kane GC, Terzic A, Puceat M (2002) Stem cell differentiation requires a paracrine pathway in the heart. Faseb J 16:1558–1566

Beltrami AP, Urbanek K, Kajstura J, Yan SM, Finato N, Bussani R, Nadal-Ginard B, Silvestri F, Leri A, Beltrami CA, Anversa P (2001) Evidence that human cardiac myocytes divide after myocardial infarction. N Engl J Med 344:1750–1757

Beltrami AP, Barlucchi L, Torella D, Baker M, Limana F, Chimenti S, Kasahara H, Rota M, Musso E, Urbanek K, Leri A, Kajstura J, Nadal-Ginard B, Anversa P (2003) Adult cardiac stem cells are multipotent and support myocardial regeneration. Cell 114:763–776

Black BL, Olson EN (1998) Transcriptional control of muscle development by myocyte enhancer factor-2 (MEF2) proteins. Annu Rev Cell Dev Biol 14:167–196

Blatt A, Robinson D, Cotter G, Efrati S, Simantov Y, Bar I, Kaluski E, Krakover R, Sidenko S, Evron Z, Lipa L, Posternak N, Nevo Z, Vered Z (2003) Improved regional left ventricular function after successful satellite cell grafting in rabbits with myocardial infarction. Eur J Heart Fail 5:751–757

Bloch W, Fleischmann BK, Lorke DE, Andressen C, Hops B, Hescheler J, Addicks K (1999) Nitric oxide synthase expression and role during cardiomyogenesis. Cardiovasc Res 43:675–684

Boheler KR, Czyz J, Tweedie D, Yang HT, Anisimov SV, Wobus AM (2002) Differentiation of pluripotent embryonic stem cells into cardiomyocytes. Circ Res 91:189–201

Boheler KR, Crider DG, Tarasova Y, Maltsev VA (2005) Cardiomyocytes derived from embryonic stem cells. Methods Mol Med 108:417–435

Bost F, Caron L, Marchetti I, Dani C, Le Marchand-Brustel Y, Binetruy B (2002) Retinoic acid activation of the ERK pathway is required for embryonic stem cell commitment into the adipocyte lineage. Biochem J 361:621–627

Brand T (2003) Heart development: molecular insights into cardiac specification and early morphogenesis. Dev Biol 258:1–19

Cai CL, Liang X, Shi Y, Chu PH, Pfaff SL, Chen J, Evans S (2003) Isl1 identifies a cardiac progenitor population that proliferates prior to differentiation and contributes a majority of cells to the heart. Dev Cell 5:877–889

Camargo FD, Chambers SM, Goodell MA (2004) Stem cell plasticity: from transdifferentiation to macrophage fusion. Cell Prolif 37:55–65

Cao F, Niu LL, Meng L, Zhao LX, Zheng M, Yue W, Bai CX, Jia GL, Pei XT (2004) [Cardiomyocyte-like differentiation of human bone marrow mesenchymal stem cells after exposure of 5-azacytidine in vitro]. Shi Yan Sheng Wu Xue Bao 37:118–124

Chen SL, Fang WW, Ye F, Liu YH, Qian J, Shan SJ, Zhang JJ, Chunhua RZ, Liao LM, Lin S, Sun JP (2004) Effect on left ventricular function of intracoronary transplantation of autologous bone marrow mesenchymal stem cell in patients with acute myocardial infarction. Am J Cardiol 94:92–95

Choi SC, Yoon JH, Shim WJ, Ro YM, Lim DS (2004) 5-azacytidine induces cardiac differentiation of P19 embryonic stem cells. Exp Mol Med 36:515–523

Dani C, Smith AG, Dessolin S, Leroy P, Staccini L, Villageois P, Darimont C, Ailhaud G (1997) Differentiation of embryonic stem cells into adipocytes in vitro. J Cell Sci 110 (Pt 11):1279–1285

Davidson SM, Morange M (2000) Hsp25 and the p38 MAPK pathway are involved in differentiation of cardiomyocytes. Dev Biol 218:146–160

Dell'Era P, Ronca R, Coco L, Nicoli S, Metra M, Presta M (2003) Fibroblast growth factor receptor-1 is essential for in vitro cardiomyocyte development. Circ Res 93:414–420

Dennis JE, Carbillet JP, Caplan AI, Charbord P (2002) The STRO-1+ marrow cell population is multipotential. Cells Tissues Organs 170:73–82

Deten A, Volz HC, Clamors S, Leiblein S, Briest W, Marx G, Zimmer HG (2005) Hematopoietic stem cells do not repair the infarcted mouse heart. Cardiovasc Res 65:52–63

Dinsmore J, Ratliff J, Deacon T, Pakzaban P, Jacoby D, Galpern W, Isacson O (1996) Embryonic stem cells differentiated in vitro as a novel source of cells for transplantation. Cell Transplant 5:131–143

Doetschman TC, Eistetter H, Katz M, Schmidt W, Kemler R (1985) The in vitro development of blastocyst-derived embryonic stem cell lines: formation of visceral yolk sac, blood islands and myocardium. J Embryol Exp Morphol 87:27–45

Doevendans PA, Kubalak SW, An RH, Becker DK, Chien KR, Kass RS (2000) Differentiation of cardiomyocytes in floating embryoid bodies is comparable to fetal cardiomyocytes. J Mol Cell Cardiol 32:839–851

Dono R, Scalera L, Pacifico F, Acampora D, Persico MG, Simeone A (1993) The murine cripto gene: expression during mesoderm induction and early heart morphogenesis. Development 118:1157–1168

Dorfman J, Duong M, Zibaitis A, Pelletier MP, Shum-Tim D, Li C, Chiu RC (1998) Myocardial tissue engineering with autologous myoblast implantation. J Thorac Cardiovasc Surg 116:744–751

Draper JS, Smith K, Gokhale P, Moore HD, Maltby E, Johnson J, Meisner L, Zwaka TP, Thomson JA, Andrews PW (2004) Recurrent gain of chromosomes 17q and 12 in cultured human embryonic stem cells. Nat Biotechnol 22:53–54

Durocher D, Charron F, Warren R, Schwartz RJ, Nemer M (1997) The cardiac transcription factors Nkx2–5 and GATA-4 are mutual cofactors. EMBO J 16:5687–5696

Edwards MK, McBurney MW (1983) The concentration of retinoic acid determines the differentiated cell types formed by a teratocarcinoma cell line. Dev Biol 98:187–191

Edwards MK, Harris JF, McBurney MW (1983) Induced muscle differentiation in an embryonal carcinoma cell line. Mol Cell Biol 3:2280–2286

Eisenberg CA, Gourdie RG, Eisenberg LM (1997) Wnt-11 is expressed in early avian mesoderm and required for the differentiation of the quail mesoderm cell line QCE-6. Development 124:525–536

Eisenberg CA, Eisenberg LM (1999) WNT11 promotes cardiac tissue formation of early mesoderm. Dev Dyn 216:45–58

Eisenberg LM, Burns L, Eisenberg CA (2003) Hematopoietic cells from bone marrow have the potential to differentiate into cardiomyocytes in vitro. Anat Rec A Discov Mol Cell Evol Biol 274:870–882

Eriksson M, Leppa S (2002) Mitogen-activated protein kinases and activator protein 1 are required for proliferation and cardiomyocyte differentiation of P19 embryonal carcinoma cells. J Biol Chem 277:15992–16001

Evans MJ, Kaufman MH (1981) Establishment in culture of pluripotential cells from mouse embryos. Nature 292:154–156

Evans SM (1999) Vertebrate tinman homologues and cardiac differentiation. Semin Cell Dev Biol 10:73–83

Fijnvandraat AC, Lekanne Deprez RH, Christoffels VM, Ruijter JM, Moorman AF (2003) TBX5 overexpression stimulates differentiation of chamber myocardium in P19C16 embryonic carcinoma cells. J Muscle Res Cell Motil 24:211–218

Fishman MC, Chien KR (1997) Fashioning the vertebrate heart: earliest embryonic decisions. Development 124:2099–2117

Foley A, Mercola M (2004) Heart induction: embryology to cardiomyocyte regeneration. Trends Cardiovasc Med 14:121–125

Foley AC, Mercola M (2005) Heart induction by Wnt antagonists depends on the homeodomain transcription factor Hex. Genes Dev 19:387–396

Foshay K, Rodriguez G, Hoel B, Narayan J, Gallicano GI (2005) JAK2/STAT3 directs cardiomyogenesis within murine embryonic stem cells in vitro. Stem Cells 23:530–543

Friedenstein AJ, Gorskaja JF, Kulagina NN (1976) Fibroblast precursors in normal and irradiated mouse hematopoietic organs. Exp Hematol 4:267–274

Gadue P, Huber TL, Nostro MC, Kattman S, Keller GM (2005) Germ layer induction from embryonic stem cells. Exp Hematol 33:955–964

Gassanov N, Er F, Zagidullin N, Hoppe UC (2004) Endothelin induces differentiation of ANP-EGFP expressing embryonic stem cells towards a pacemaker phenotype. FASEB J 18:1710–1712

Ghosh-Choudhury N, Abboud SL, Mahimainathan L, Chandrasekar B, Choudhury GG (2003) Phosphatidylinositol 3-kinase regulates bone morphogenetic protein-2 (BMP-2)-induced myocyte enhancer factor 2A-dependent transcription of BMP-2 gene in cardiomyocyte precursor cells. J Biol Chem 278:21998–2005

Gianakopoulos PJ, Skerjanc IS (2005) Hedgehog signaling induces cardiomyogenesis in P19 cells. J Biol Chem 280:21022–21028

Gnecchi M, He H, Liang OD, Melo LG, Morello F, Mu H, Noiseux N, Zhang L, Pratt RE, Ingwall JS, Dzau VJ (2005) Paracrine action accounts for marked protection of ischemic heart by Akt-modified mesenchymal stem cells. Nat Med 11:367–368

Gojo S, Gojo N, Takeda Y, Mori T, Abe H, Kyo S, Hata J, Umezawa A (2003) In vivo cardiovasculogenesis by direct injection of isolated adult mesenchymal stem cells. Exp Cell Res 288:51–59

Grepin C, Nemer G, Nemer M (1997) Enhanced cardiogenesis in embryonic stem cells overexpressing the GATA-4 transcription factor. Development 124:2387–2395

Guan K, Furst DO, Wobus AM (1999) Modulation of sarcomere organization during embryonic stem cell-derived cardiomyocyte differentiation. Eur J Cell Biol 78:813–823

Habara-Ohkubo A (1996) Differentiation of beating cardiac muscle cells from a derivative of P19 embryonal carcinoma cells. Cell Struct Funct 21:101–110

Hakuno D, Fukuda K, Makino S, Konishi F, Tomita Y, Manabe T, Suzuki Y, Umezawa A, Ogawa S (2002) Bone marrow-derived regenerated cardiomyocytes (CMG cells) express functional adrenergic and muscarinic receptors. Circulation 105:380–386

Haynesworth SE, Baber MA, Caplan AI (1992) Cell surface antigens on human marrow-derived mesenchymal cells are detected by monoclonal antibodies. Bone 13:69–80

He JQ, Ma Y, Lee Y, Thomson JA, Kamp TJ (2003) Human embryonic stem cells develop into multiple types of cardiac myocytes: action potential characterization. Circ Res 93:32–39

Hidai C, Masako O, Ikeda H, Nagashima H, Matsuoka R, Quertermous T, Kasanuki H, Kokubun S, Kawana M (2003) FGF-1 enhanced cardiogenesis in differentiating embryonal carcinoma cell cultures, which was opposite to the effect of FGF-2. J Mol Cell Cardiol 35:421–425

Hierlihy AM, Seale P, Lobe CG, Rudnicki MA, Megeney LA (2002) The post-natal heart contains a myocardial stem cell population. FEBS Lett 530:239–243

Hiroi Y, Kudoh S, Monzen K, Ikeda Y, Yazaki Y, Nagai R, Komuro I (2001) Tbx5 associates with Nkx2–5 and synergistically promotes cardiomyocyte differentiation. Nat Genet 28:276–280

Hoffman LM, Carpenter MK (2005) Characterization and culture of human embryonic stem cells. Nat Biotechnol 23:699–708

Horackova M, Arora R, Chen R, Armour JA, Cattini PA, Livingston R, Byczko Z (2004) Cell transplantation for treatment of acute myocardial infarction: unique capacity for repair by skeletal muscle satellite cells. Am J Physiol Heart Circ Physiol 287:H1599–1608

Hosoda T, Monzen K, Hiroi Y, Oka T, Takimoto E, Yazaki Y, Nagai R, Komuro I (2001) A novel myocyte-specific gene Midori promotes the differentiation of P19CL6 cells into cardiomyocytes. J Biol Chem 276:35978–35989

Iida M, Heike T, Yoshimoto M, Baba S, Doi H, Nakahata T (2005) Identification of cardiac stem cells with FLK1, CD31, and VE-cadherin expression during embryonic stem cell differentiation. FASEB J 19:371–378

In 't Anker PS, Scherjon SA, Kleijburg-van der Keur C, de Groot-Swings GM, Claas FH, Fibbe WE, Kanhai HH (2004) Isolation of mesenchymal stem cells of fetal or maternal origin from human placenta. Stem Cells 22:1338–1345

Itskovitz-Eldor J, Schuldiner M, Karsenti D, Eden A, Yanuka O, Amit M, Soreq H, Benvenisty N (2000) Differentiation of human embryonic stem cells into embryoid bodies compromising the three embryonic germ layers. Mol Med 6:88–95

Jackson KA, Majka SM, Wang H, Pocius J, Hartley CJ, Majesky MW, Entman ML, Michael LH, Hirschi KK, Goodell MA (2001) Regeneration of ischemic cardiac muscle and vascular endothelium by adult stem cells. J Clin Invest 107:1395–1402

Jiang Y, Jahagirdar BN, Reinhardt RL, Schwartz RE, Keene CD, Ortiz-Gonzalez XR, Reyes M, Lenvik T, Lund T, Blackstad M, Du J, Aldrich S, Lisberg A, Low WC, Largaespada DA, Verfaillie CM (2002) Pluripotency of mesenchymal stem cells derived from adult marrow. Nature 418:41–49

Kajstura J, Rota M, Whang B, Cascapera S, Hosoda T, Bearzi C, Nurzynska D, Kasahara H, Zias E, Bonafe M, Nadal-Ginard B, Torella D, Nascimbene A, Quaini F, Urbanek K, Leri A, Anversa P (2005) Bone marrow cells differentiate in cardiac cell lineages after infarction independently of cell fusion. Circ Res 96:127–137

Kanno S, Kim PK, Sallam K, Lei J, Billiar TR, Shears LL 2nd (2004) Nitric oxide facilitates cardiomyogenesis in mouse embryonic stem cells. Proc Natl Acad Sci USA 101:12277–12281

Kehat I, Kenyagin-Karsenti D, Snir M, Segev H, Amit M, Gepstein A, Livne E, Binah O, Itskovitz-Eldor J, Gepstein L (2001) Human embryonic stem cells can differentiate into myocytes with structural and functional properties of cardiomyocytes. J Clin Invest 108:407–414

Kehat I, Gepstein A, Spira A, Itskovitz-Eldor J, Gepstein L (2002) High-resolution electrophysiological assessment of human embryonic stem cell-derived cardiomyocytes: a novel in vitro model for the study of conduction. Circ Res 91:659–661

Kehat I, Khimovich L, Caspi O, Gepstein A, Shofti R, Arbel G, Huber I, Satin J, Itskovitz-Eldor J, Gepstein L (2004) Electromechanical integration of cardiomyocytes derived from human embryonic stem cells. Nat Biotechnol 22:1282–1289

Keller G (2005) Embryonic stem cell differentiation: emergence of a new era in biology and medicine. Genes Dev 19:1129–1155

Kleinsmith LJ, Pierce GB Jr (1964) Multipotentiality of single embryonal carcinoma cells. Cancer Res 24:1544–1551

Klinz F, Bloch W, Addicks K, Hescheler J (1999) Inhibition of phosphatidylinositol-3-kinase blocks development of functional embryonic cardiomyocytes. Exp Cell Res 247:79–83

Klug MG, Soonpaa MH, Koh GY, Field LJ (1996) Genetically selected cardiomyocytes from differentiating embryonic stem cells form stable intracardiac grafts. J Clin Invest 98:216–224

Kogler G, Sensken S, Airey JA, Trapp T, Muschen M, Feldhahn N, Liedtke S, Sorg RV, Fischer J, Rosenbaum C, Greschat S, Knipper A, Bender J, Degistirici O, Gao J, Caplan AI, Colletti EJ, Almeida-Porada G, Muller HW, Zanjani E, Wernet P (2004) A new human somatic stem cell from placental cord blood with intrinsic pluripotent differentiation potential. J Exp Med 200:123–135

Kolossov E, Lu Z, Drobinskaya I, Gassanov N, Duan Y, Sauer H, Manzke O, Bloch W, Bohlen H, Hescheler J, Fleischmann BK (2005) Identification and characterization of embryonic stem cell-derived pacemaker and atrial cardiomyocytes. Faseb J 19:577–579

Kouskoff V, Lacaud G, Schwantz S, Fehling HJ, Keller G (2005) Sequential development of hematopoietic and cardiac mesoderm during embryonic stem cell differentiation. Proc Natl Acad Sci USA 102:13170–13175

Kumar D, Sun B (2005) Transforming growth factor-beta2 enhances differentiation of cardiac myocytes from embryonic stem cells. Biochem Biophys Res Commun 332:135–141

Lakshmipathy U, Verfaillie C (2005) Stem cell plasticity. Blood Rev 19:29–38

Lapidos KA, Chen YE, Earley JU, Heydemann A, Huber JM, Chien M, Ma A, McNally EM (2004) Transplanted hematopoietic stem cells demonstrate impaired sarcoglycan expression after engraftment into cardiac and skeletal muscle. J Clin Invest 114:1577–1585

Laugwitz KL, Moretti A, Lam J, Gruber P, Chen Y, Woodard S, Lin LZ, Cai CL, Lu MM, Reth M, Platoshyn O, Yuan JX, Evans S, Chien KR (2005) Postnatal isl1+ cardioblasts enter fully differentiated cardiomyocyte lineages. Nature 433:647–653

Lawrenz B, Schiller H, Willbold E, Ruediger M, Muhs A, Esser S (2004) Highly sensitive biosafety model for stem-cell-derived grafts. Cytotherapy 6:212–222

Leobon B, Garcin I, Menasche P, Vilquin JT, Audinat E, Charpak S (2003) Myoblasts transplanted into rat infarcted myocardium are functionally isolated from their host. Proc Natl Acad Sci USA 100:7808–7811

Li J, Puceat M, Perez-Terzic C, Mery A, Nakamura K, Michalak M, Krause KH, Jaconi ME (2002) Calreticulin reveals a critical Ca(2+) checkpoint in cardiac myofibrillogenesis. J Cell Biol 158:103–113

Lickert H, Kutsch S, Kanzler B, Tamai Y, Taketo MM, Kemler R (2002) Formation of multiple hearts in mice following deletion of beta-catenin in the embryonic endoderm. Dev Cell 3:171–181

Liu H, Harris TM, Kim HH, Childs G (2005) Cardiac myocyte differentiation: the Nkx2.5 and Cripto target genes in P19 clone 6 cells. Funct Integr Genomics 5:218–239

Liu Y, Song J, Liu W, Wan Y, Chen X, Hu C (2003) Growth and differentiation of rat bone marrow stromal cells: does 5-azacytidine trigger their cardiomyogenic differentiation? Cardiovasc Res 58:460–468

Logan M, Mohun T (1993) Induction of cardiac muscle differentiation in isolated animal pole explants of Xenopus laevis embryos. Development 118:865–875

Lough J, Sugi Y (2000) Endoderm and heart development. Dev Dyn 217:327–342

Lough J, Barron M, Brogley M, Sugi Y, Bolender DL, Zhu X (1996) Combined BMP-2 and FGF-4, but neither factor alone, induces cardiogenesis in non-precardiac embryonic mesoderm. Dev Biol 178:198–202

Makino S, Fukuda K, Miyoshi S, Konishi F, Kodama H, Pan J, Sano M, Takahashi T, Hori S, Abe H, Hata J, Umezawa A, Ogawa S (1999) Cardiomyocytes can be generated from marrow stromal cells in vitro. J Clin Invest 103:697–705

Maltsev VA, Rohwedel J, Hescheler J, Wobus AM (1993) Embryonic stem cells differentiate in vitro into cardiomyocytes representing sinusnodal, atrial and ventricular cell types. Mech Dev 44:41–50

Maltsev VA, Wobus AM, Rohwedel J, Bader M, Hescheler J (1994) Cardiomyocytes differentiated in vitro from embryonic stem cells developmentally express cardiac-specific genes and ionic currents. Circ Res 75:233–244

Manabe K, Miake J, Sasaki N, Furuichi H, Yano S, Mizuta E, Yamamoto Y, Hoshikawa Y, Yamazaki H, Tajima F, Shiota G, Nanba E, Ohgi S, Hidaka K, Morisaki T, Kurata Y, Lee JK, Igawa O, Shigemasa C, Hisatome I (2004) Developmental changes of Ni(2+) sensitivity and automaticity in Nkx2.5-positive cardiac precursor cells from murine embryonic stem cell. Circ J 68:724–726

Mangi AA, Noiseux N, Kong D, He H, Rezvani M, Ingwall JS, Dzau VJ (2003) Mesenchymal stem cells modified with Akt prevent remodeling and restore performance of infarcted hearts. Nat Med 9:1195–1201

Martin GR, Evans MJ (1974) The morphology and growth of a pluripotent teratocarcinoma cell line and its derivatives in tissue culture. Cell 2:163–172

Martin GR, Evans MJ (1975) Differentiation of clonal lines of teratocarcinoma cells: formation of embryoid bodies in vitro. Proc Natl Acad Sci USA 72:1441–1445

Matsuura K, Wada H, Nagai T, Iijima Y, Minamino T, Sano M, Akazawa H, Molkentin JD, Kasanuki H, Komuro I (2004) Cardiomyocytes fuse with surrounding noncardiomyocytes and reenter the cell cycle. J Cell Biol 167:351–363

McBurney MW, Rogers BJ (1982) Isolation of male embryonal carcinoma cells and their chromosome replication patterns. Dev Biol 89:503–508

McBurney MW, Jones-Villeneuve EM, Edwards MK, Anderson PJ (1982) Control of muscle and neuronal differentiation in a cultured embryonal carcinoma cell line. Nature 299:165–167

Menasche P, Hagege AA, Scorsin M, Pouzet B, Desnos M, Duboc D, Schwartz K, Vilquin JT, Marolleau JP (2001) Myoblast transplantation for heart failure. Lancet 357:279–280

Menasche P, Hagege AA, Vilquin JT, Desnos M, Abergel E, Pouzet B, Bel A, Sarateanu S, Scorsin M, Schwartz K, Bruneval P, Benbunan M, Marolleau JP, Duboc D (2003) Autologous skeletal myoblast transplantation for severe postinfarction left ventricular dysfunction. J Am Coll Cardiol 41:1078–1083

Messina E, De Angelis L, Frati G, Morrone S, Chimenti S, Fiordaliso F, Salio M, Battaglia M, Latronico MV, Coletta M, Vivarelli E, Frati L, Cossu G, Giacomello A (2004) Isolation and expansion of adult cardiac stem cells from human and murine heart. Circ Res 95:911–921

Metzger JM, Lin WI, Samuelson LC (1994) Transition in cardiac contractile sensitivity to calcium during the in vitro differentiation of mouse embryonic stem cells. J Cell Biol 126:701–711

Meyer N, Jaconi M, Landopoulou A, Fort P, Puceat M (2000) A fluorescent reporter gene as a marker for ventricular specification in ES-derived cardiac cells. FEBS Lett 478:151–158

Miller-Hance WC, LaCorbiere M, Fuller SJ, Evans SM, Lyons G, Schmidt C, Robbins J, Chien KR (1993) In vitro chamber specification during embryonic stem cell cardiogenesis. Expression of the ventricular myosin light chain-2 gene is independent of heart tube formation. J Biol Chem 268:25244–25252

Molkentin JD (2000) The zinc finger-containing transcription factors GATA-4, -5, and -6. Ubiquitously expressed regulators of tissue-specific gene expression. J Biol Chem 275:38949–38952

Monzen K, Shiojima I, Hiroi Y, Kudoh S, Oka T, Takimoto E, Hayashi D, Hosoda T, Habara-Ohkubo A, Nakaoka T, Fujita T, Yazaki Y, Komuro I (1999) Bone morphogenetic proteins induce cardiomyocyte differentiation through the mitogen-activated protein kinase kinase kinase TAK1 and cardiac transcription factors Csx/Nkx-2.5 and GATA-4. Mol Cell Biol 19:7096–7105

Monzen K, Hiroi Y, Kudoh S, Akazawa H, Oka T, Takimoto E, Hayashi D, Hosoda T, Kawabata M, Miyazono K, Ishii S, Yazaki Y, Nagai R, Komuro I (2001) Smads, TAK1, and their common target ATF-2 play a critical role in cardiomyocyte differentiation. J Cell Biol 153:687–698

Monzen K, Zhu W, Kasai H, Hiroi Y, Hosoda T, Akazawa H, Zou Y, Hayashi D, Yamazaki T, Nagai R, Komuro I (2002) Dual effects of the homeobox transcription factor Csx/Nkx2–5 on cardiomyocytes. Biochem Biophys Res Commun 298:493–500

Morin S, Charron F, Robitaille L, Nemer M (2000) GATA-dependent recruitment of MEF2 proteins to target promoters. EMBO J 19:2046–2055

Mummery C, Ward D, van den Brink CE, Bird SD, Doevendans PA, Opthof T, Brutel de la Riviere A, Tertoolen L, van der Heyden M, Pera M (2002) Cardiomyocyte differentiation of mouse and human embryonic stem cells. J Anat 200:233–242

Mummery C, Ward-van Oostwaard D, Doevendans P, Spijker R, van den Brink S, Hassink R, van der Heyden M, Opthof T, Pera M, de la Riviere AB, Passier R, Tertoolen L (2003) Differentiation of human embryonic stem cells to cardiomyocytes: role of coculture with visceral endoderm-like cells. Circulation 107:2733–2740

Mummery CL, van Achterberg TA, van den Eijnden-van Raaij AJ, van Haaster L, Willemse A, de Laat SW, Piersma AH (1991) Visceral-endoderm-like cell lines induce differentiation of murine P19 embryonal carcinoma cells. Differentiation 46:51–60

Murry CE, Soonpaa MH, Reinecke H, Nakajima H, Nakajima HO, Rubart M, Pasumarthi KB, Virag JI, Bartelmez SH, Poppa V, Bradford G, Dowell JD, Williams DA, Field LJ (2004) Haematopoietic stem cells do not transdifferentiate into cardiac myocytes in myocardial infarcts. Nature 428:664–668

Muthuchamy M, Pajak L, Howles P, Doetschman T, Wieczorek DF (1993) Developmental analysis of tropomyosin gene expression in embryonic stem cells and mouse embryos. Mol Cell Biol 13:3311–3323

Nagaya N, Fujii T, Iwase T, Ohgushi H, Itoh T, Uematsu M, Yamagishi M, Mori H, Kangawa K, Kitamura S (2004) Intravenous administration of mesenchymal stem cells improves cardiac function in rats with acute myocardial infarction through angiogenesis and myogenesis. Am J Physiol Heart Circ Physiol 287:H2670–H2676

Naito AT, Tominaga A, Oyamada M, Oyamada Y, Shiraishi I, Monzen K, Komuro I, Takamatsu T (2003) Early stage-specific inhibitions of cardiomyocyte differentiation and expression of Csx/Nkx-2.5 and GATA-4 by phosphatidylinositol 3-kinase inhibitor LY294002. Exp Cell Res 291:56–69

Nakamura T, Sano M, Songyang Z, Schneider MD (2003) A Wnt- and beta -catenin-dependent pathway for mammalian cardiac myogenesis. Proc Natl Acad Sci USA 100:5834–5839

Ng WA, Doetschman T, Robbins J, Lessard JL (1997) Muscle isoactin expression during in vitro differentiation of murine embryonic stem cells. Pediatr Res 41:285–292

Nygren JM, Jovinge S, Breitbach M, Sawen P, Roll W, Hescheler J, Taneera J, Fleischmann BK, Jacobsen SE (2004) Bone marrow-derived hematopoietic cells generate cardiomyocytes at a low frequency through cell fusion, but not transdifferentiation. Nat Med 10:494–501

Oh H, Bradfute SB, Gallardo TD, Nakamura T, Gaussin V, Mishina Y, Pocius J, Michael LH, Behringer RR, Garry DJ, Entman ML, Schneider MD (2003) Cardiac progenitor cells from adult myocardium: homing, differentiation, and fusion after infarction. Proc Natl Acad Sci USA 100:12313–12318

Oh H, Chi X, Bradfute SB, Mishina Y, Pocius J, Michael LH, Behringer RR, Schwartz RJ, Entman ML, Schneider MD (2004) Cardiac muscle plasticity in adult and embryo by heart-derived progenitor cells. Ann N Y Acad Sci 1015:182–189

Orlic D, Kajstura J, Chimenti S, Bodine DM, Leri A, Anversa P (2001a) Transplanted adult bone marrow cells repair myocardial infarcts in mice. Ann N Y Acad Sci 938:221–229; discussion 229–230

Orlic D, Kajstura J, Chimenti S, Jakoniuk I, Anderson SM, Li B, Pickel J, McKay R, Nadal-Ginard B, Bodine DM, Leri A, Anversa P (2001b) Bone marrow cells regenerate infarcted myocardium. Nature 410:701–705

Orlic D, Kajstura J, Chimenti S, Limana F, Jakoniuk I, Quaini F, Nadal-Ginard B, Bodine DM, Leri A, Anversa P (2001c) Mobilized bone marrow cells repair the infarcted heart, improving function and survival. Proc Natl Acad Sci USA 98:10344–10349

Otsu K, Kuruma A, Yanagida E, Shoji S, Inoue T, Hirayama Y, Uematsu H, Hara Y, Kawano S (2005) Na+/K+ ATPase and its functional coupling with Na+/Ca2+ exchanger in mouse embryonic stem cells during differentiation into cardiomyocytes. Cell Calcium 37:137–151
Pandur P, Lasche M, Eisenberg LM, Kuhl M (2002) Wnt-11 activation of a non-canonical Wnt signalling pathway is required for cardiogenesis. Nature 418:636–641
Paquin J, Danalache BA, Jankowski M, McCann SM, Gutkowska J (2002) Oxytocin induces differentiation of P19 embryonic stem cells to cardiomyocytes. Proc Natl Acad Sci USA 99:9550–9555
Parisi S, D'Andrea D, Lago CT, Adamson ED, Persico MG, Minchiotti G (2003) Nodal-dependent Cripto signaling promotes cardiomyogenesis and redirects the neural fate of embryonic stem cells. J Cell Biol 163:303–314
Passier R, Oostwaard DW, Snapper J, Kloots J, Hassink RJ, Kuijk E, Roelen B, de la Riviere AB, Mummery C (2005) Increased cardiomyocyte differentiation from human embryonic stem cells in serum-free cultures. Stem Cells 23:772–780
Peng CF, Wei Y, Levsky JM, McDonald TV, Childs G, Kitsis RN (2002) Microarray analysis of global changes in gene expression during cardiac myocyte differentiation. Physiol Genomics 9:145–155
Pereira RF, Halford KW, O'Hara MD, Leeper DB, Sokolov BP, Pollard MD, Bagasra O, Prockop DJ (1995) Cultured adherent cells from marrow can serve as long-lasting precursor cells for bone, cartilage, and lung in irradiated mice. Proc Natl Acad Sci USA 92:4857–4861
Pittenger MF, Mackay AM, Beck SC, Jaiswal RK, Douglas R, Mosca JD, Moorman MA, Simonetti DW, Craig S, Marshak DR (1999) Multilineage potential of adult human mesenchymal stem cells. Science 284:143–147
Potapova I, Plotnikov A, Lu Z, Danilo P Jr, Valiunas V, Qu J, Doronin S, Zuckerman J, Shlapakova IN, Gao J, Pan Z, Herron AJ, Robinson RB, Brink PR, Rosen MR, Cohen IS (2004) Human mesenchymal stem cells as a gene delivery system to create cardiac pacemakers. Circ Res 94:952–959
Puceat M, Jaconi M (2005) Ca(2+) signalling in cardiogenesis. Cell Calcium 38:383–389
Qu CK, Feng GS (1998) Shp-2 has a positive regulatory role in ES cell differentiation and proliferation. Oncogene 17:433–439
Quesenberry PJ, Dooner G, Colvin G, Abedi M (2005) Stem cell biology and the plasticity polemic. Exp Hematol 33:389–394
Rastan AJ, Walther T, Kostelka M, Garbade J, Schubert A, Stein A, Dhein S, Mohr FW (2005) Morphological, electrophysiological and coupling characteristics of bone marrow-derived mononuclear cells-an in vitro-model. Eur J Cardiothorac Surg 27:104–110
Reinecke H, Poppa V, Murry CE (2002) Skeletal muscle stem cells do not transdifferentiate into cardiomyocytes after cardiac grafting. J Mol Cell Cardiol 34:241–249
Reiss K, Kajstura J, Zhang X, Li P, Szoke E, Olivetti G, Anversa P (1994) Acute myocardial infarction leads to upregulation of the IGF-1 autocrine system, DNA replication, and nuclear mitotic division in the remaining viable cardiac myocytes. Exp Cell Res 213:463–472
Reppel M, Boettinger C, Hescheler J (2004) Beta-adrenergic and muscarinic modulation of human embryonic stem cell-derived cardiomyocytes. Cell Physiol Biochem 14:187–196
Reyes M, Verfaillie CM (2001) Characterization of multipotent adult progenitor cells, a subpopulation of mesenchymal stem cells. Ann NY Acad Sci 938:231–233; discussion 233–235
Robbins J, Gulick J, Sanchez A, Howles P, Doetschman T (1990) Mouse embryonic stem cells express the cardiac myosin heavy chain genes during development in vitro. J Biol Chem 265:11905–11909
Rodriguez ER, Tan CD, Onwuta US, Yu ZX, Ferrans VJ, Parrillo JE (1994) 3,5,3'-Triiodo-L-thyronine induces cardiac myocyte differentiation but not neuronal differentiation in P19 teratocarcinoma cells in a dose dependent manner. Biochem Biophys Res Commun 205:652–658
Romanov YA, Svintsitskaya VA, Smirnov VN (2003) Searching for alternative sources of postnatal human mesenchymal stem cells: candidate MSC-like cells from umbilical cord. Stem Cells 21:105–110
Rosenblatt-Velin N, Lepore MG, Cartoni C, Beermann F, Pedrazzini T (2005) FGF-2 controls the differentiation of resident cardiac precursors into functional cardiomyocytes. J Clin Invest 115:1724–1733
Rubart M, Soonpaa MH, Nakajima H, Field LJ (2004) Spontaneous and evoked intracellular calcium transients in donor-derived myocytes following intracardiac myoblast transplantation. J Clin Invest 114:775–783
Rudnicki MA, Jackowski G, Saggin L, McBurney MW (1990) Actin and myosin expression during development of cardiac muscle from cultured embryonal carcinoma cells. Dev Biol 138:348–358
Rudy-Reil D, Lough J (2004) Avian precardiac endoderm/mesoderm induces cardiac myocyte differentiation in murine embryonic stem cells. Circ Res 94:e107–e116
Ryan K, Chin AJ (2003) T-box genes and cardiac development. Birth Defects Res C Embryo Today 69:25–37

Sachinidis A, Gissel C, Nierhoff D, Hippler-Altenburg R, Sauer H, Wartenberg M, Hescheler J (2003) Identification of platelet-derived growth factor-BB as cardiogenesis-inducing factor in mouse embryonic stem cells under serum-free conditions. Cell Physiol Biochem 13:423–429

Sanchez A, Jones WK, Gulick J, Doetschman T, Robbins J (1991) Myosin heavy chain gene expression in mouse embryoid bodies. An in vitro developmental study. J Biol Chem 266:22419–22426

Satin J, Kehat I, Caspi O, Huber I, Arbel G, Itzhaki I, Magyar J, Schroder EA, Perlman I, Gepstein L (2004) Mechanism of spontaneous excitability in human embryonic stem cell derived cardiomyocytes. J Physiol 559:479–496

Sauer H, Rahimi G, Hescheler J, Wartenberg M (2000) Role of reactive oxygen species and phosphatidylinositol 3-kinase in cardiomyocyte differentiation of embryonic stem cells. FEBS Lett 476:218–223

Sauer H, Neukirchen W, Rahimi G, Grunheck F, Hescheler J, Wartenberg M (2004) Involvement of reactive oxygen species in cardiotrophin-1-induced proliferation of cardiomyocytes differentiated from murine embryonic stem cells. Exp Cell Res 294:313–324

Schier AF, Shen MM (2000) Nodal signalling in vertebrate development. Nature 403:385–389

Schuldiner M, Yanuka O, Itskovitz-Eldor J, Melton DA, Benvenisty N (2000) Effects of eight growth factors on the differentiation of cells derived from human embryonic stem cells. Proc Natl Acad Sci USA 97:11307–11312

Scorsin M, Hagege A, Vilquin JT, Fiszman M, Marotte F, Samuel JL, Rappaport L, Schwartz K, Menasche P (2000) Comparison of the effects of fetal cardiomyocyte and skeletal myoblast transplantation on postinfarction left ventricular function. J Thorac Cardiovasc Surg 119:1169–1175

Siminiak T, Kalawski R, Fiszer D, Jerzykowska O, Rzezniczak J, Rozwadowska N, Kurpisz M (2004) Autologous skeletal myoblast transplantation for the treatment of postinfarction myocardial injury: phase I clinical study with 12 months of follow-up. Am Heart J 148:531–537

Simmons PJ, Torok-Storb B (1991) Identification of stromal cell precursors in human bone marrow by a novel monoclonal antibody, STRO-1. Blood 78:55–62

Skerjanc IS, Petropoulos H, Ridgeway AG, Wilton S (1998) Myocyte enhancer factor 2C and Nkx2–5 upregulate each other's expression and initiate cardiomyogenesis in P19 cells. J Biol Chem 273:34904–34910

Smith SC, Reuhl KR, Craig J, McBurney MW (1987) The role of aggregation in embryonal carcinoma cell differentiation. J Cell Physiol 131:74–84

Snir M, Kehat I, Gepstein A, Coleman R, Itskovitz-Eldor J, Livne E, Gepstein L (2003) Assessment of the ultrastructural and proliferative properties of human embryonic stem cell-derived cardiomyocytes. Am J Physiol Heart Circ Physiol 285:H2355–H2363

Srivastava D, Olson EN (2000) A genetic blueprint for cardiac development. Nature 407:221–226

Suk Kim H, Hidaka K, Morisaki T (2003) Expression of ErbB receptors in ES cell-derived cardiomyocytes. Biochem Biophys Res Commun 309:241–246

Takahashi T, Lord B, Schulze PC, Fryer RM, Sarang SS, Gullans SR, Lee RT (2003) Ascorbic acid enhances differentiation of embryonic stem cells into cardiac myocytes. Circulation 107:1912–1916

Takeda Y, Mori T, Imabayashi H, Kiyono T, Gojo S, Miyoshi S, Hida N, Ita M, Segawa K, Ogawa S, Sakamoto M, Nakamura S, Umezawa A (2004) Can the life span of human marrow stromal cells be prolonged by bmi-1, E6, E7, and/or telomerase without affecting cardiomyogenic differentiation? J Gene Med 6:833–845

Taylor SM, Jones PA (1979) Multiple new phenotypes induced in 10T1/2 and 3T3 cells treated with 5-azacytidine. Cell 17:771–779

Terada N, Hamazaki T, Oka M, Hoki M, Mastalerz DM, Nakano Y, Meyer EM, Morel L, Petersen BE, Scott EW (2002) Bone marrow cells adopt the phenotype of other cells by spontaneous cell fusion. Nature 416:542–545

Terami H, Hidaka K, Katsumata T, Iio A, Morisaki T (2004) Wnt11 facilitates embryonic stem cell differentiation to Nkx2.5-positive cardiomyocytes. Biochem Biophys Res Commun 325:968–975

Thomson JA, Itskovitz-Eldor J, Shapiro SS, Waknitz MA, Swiergiel JJ, Marshall VS, Jones JM (1998) Embryonic stem cell lines derived from human blastocysts. Science 282:1145–1147

Toma C, Pittenger MF, Cahill KS, Byrne BJ, Kessler PD (2002) Human mesenchymal stem cells differentiate to a cardiomyocyte phenotype in the adult murine heart. Circulation 105:93–98

Urbanek K, Quaini F, Tasca G, Torella D, Castaldo C, Nadal-Ginard B, Leri A, Kajstura J, Quaini E, Anversa P (2003) Intense myocyte formation from cardiac stem cells in human cardiac hypertrophy. Proc Natl Acad Sci USA 100:10440–10445

Urbanek K, Torella D, Sheikh F, De Angelis A, Nurzynska D, Silvestri F, Beltrami CA, Bussani R, Beltrami AP, Quaini F, Bolli R, Leri A, Kajstura J, Anversa P (2005) Myocardial regeneration by activation of multipotent cardiac stem cells in ischemic heart failure. Proc Natl Acad Sci USA 102:8692–8697

Van der Heyden MA, van Kempen MJ, Tsuji Y, Rook MB, Jongsma HJ, Opthof T (2003) P19 embryonal carcinoma cells: a suitable model system for cardiac electrophysiological differentiation at the molecular and functional level. Cardiovasc Res 58:410–422

Van Tuyn J, Knaan-Shanzer S, van de Watering MJ, de Graaf M, van der Laarse A, Schalij MJ, van der Wall EE, de Vries AA, Atsma DE (2005) Activation of cardiac and smooth muscle-specific genes in primary human cells after forced expression of human myocardin. Cardiovasc Res 67:245–255

Ventura C, Maioli M (2000) Opioid peptide gene expression primes cardiogenesis in embryonal pluripotent stem cells. Circ Res 87:189–194

Ventura C, Maioli M, Asara Y, Santoni D, Scarlata I, Cantoni S, Perbellini A (2004) Butyric and retinoic mixed ester of hyaluronan. A novel differentiating glycoconjugate affording a high throughput of cardiogenesis in embryonic stem cells. J Biol Chem 279:23574–23579

Wagers AJ, Weissman IL (2004) Plasticity of adult stem cells. Cell 116:639–648

Wang Z, Cohen K, Shao Y, Mole P, Dombkowski D, Scadden DT (2004) Ephrin receptor, EphB4, regulates ES cell differentiation of primitive mammalian hemangioblasts, blood, cardiomyocytes, and blood vessels. Blood 103:100–109

Wheatley SC, Isacke CM, Crossley PH (1993) Restricted expression of the hyaluronan receptor, CD44, during postimplantation mouse embryogenesis suggests key roles in tissue formation and patterning. Development 119:295–306

Winitsky SO, Gopal TV, Hassanzadeh S, Takahashi H, Gryder D, Rogawski MA, Takeda K, Yu ZX, Xu YH, Epstein ND (2005) Adult murine skeletal muscle contains cells that can differentiate into beating cardiomyocytes in vitro. PLoS Biol 3:e87

Wobus AM, Wallukat G, Hescheler J (1991) Pluripotent mouse embryonic stem cells are able to differentiate into cardiomyocytes expressing chronotropic responses to adrenergic and cholinergic agents and Ca2+ channel blockers. Differentiation 48:173–182

Wobus AM, Kleppisch T, Maltsev V, Hescheler J (1994) Cardiomyocyte-like cells differentiated in vitro from embryonic carcinoma cells P19 are characterized by functional expression of adrenoceptors and Ca2+ channels. In Vitro Cell Dev Biol Anim 30A:425–434

Wobus AM, Kaomei G, Shan J, Wellner MC, Rohwedel J, Ji G, Fleischmann B, Katus HA, Hescheler J, Franz WM (1997) Retinoic acid accelerates embryonic stem cell-derived cardiac differentiation and enhances development of ventricular cardiomyocytes. J Mol Cell Cardiol 29:1525–1539

Wollert KC, Meyer GP, Lotz J, Ringes-Lichtenberg S, Lippolt P, Breidenbach C, Fichtner S, Korte T, Hornig B, Messinger D, Arseniev L, Hertenstein B, Ganser A, Drexler H (2004) Intracoronary autologous bone-marrow cell transfer after myocardial infarction: the BOOST randomised controlled clinical trial. Lancet 364:141–148

Wu X, Ding S, Ding Q, Gray NS, Schultz PG (2004) Small molecules that induce cardiomyogenesis in embryonic stem cells. J Am Chem Soc 126:1590–1591

Xu C, Liguori G, Adamson ED, Persico MG (1998) Specific arrest of cardiogenesis in cultured embryonic stem cells lacking Cripto-1. Dev Biol 196:237–247

Xu C, Police S, Rao N, Carpenter MK (2002) Characterization and enrichment of cardiomyocytes derived from human embryonic stem cells. Circ Res 91:501–508

Xu W, Zhang X, Qian H, Zhu W, Sun X, Hu J, Zhou H, Chen Y (2004) Mesenchymal stem cells from adult human bone marrow differentiate into a cardiomyocyte phenotype in vitro. Exp Biol Med (Maywood) 229:623–631

Xue T, Cho HC, Akar FG, Tsang SY, Jones SP, Marban E, Tomaselli GF, Li RA (2005) Functional integration of electrically active cardiac derivatives from genetically engineered human embryonic stem cells with quiescent recipient ventricular cardiomyocytes: insights into the development of cell-based pacemakers. Circulation 111:11–20

Yang Y, Min JY, Rana JS, Ke Q, Cai J, Chen Y, Morgan JP, Xiao YF (2002) VEGF enhances functional improvement of postinfarcted hearts by transplantation of ESC-differentiated cells. J Appl Physiol 93:1140–1151

Yoon YS, Wecker A, Heyd L, Park JS, Tkebuchava T, Kusano K, Hanley A, Scadova H, Qin G, Cha DH, Johnson KL, Aikawa R, Asahara T, Losordo DW (2005) Clonally expanded novel multipotent stem cells from human bone marrow regenerate myocardium after myocardial infarction. J Clin Invest 115:326–338

Yuasa S, Itabashi Y, Koshimizu U, Tanaka T, Sugimura K, Kinoshita M, Hattori F, Fukami SI, Shimazaki T, Okano H, Ogawa S, Fukuda K (2005) Transient inhibition of BMP signaling by Noggin induces cardiomyocyte differentiation of mouse embryonic stem cells. Nat Biotechnol 23:607–611

Zhang F, Chen Y, Yang Z, Gao X, Ma W, Li C, Kao RL (2003) Cellular cardiomyoplasty for a patient with heart failure. Cardiovasc Radiat Med 4:43–46

Zhang S, Wang D, Estrov Z, Raj S, Willerson JT, Yeh ET (2004) Both cell fusion and transdifferentiation account for the transformation of human peripheral blood CD34-positive cells into cardiomyocytes in vivo. Circulation 110:3803–3807

Zhang YM, Shang L, Hartzell C, Narlow M, Cribbs L, Dudley SC Jr (2003) Characterization and regulation of T-type Ca2+ channels in embryonic stem cell-derived cardiomyocytes. Am J Physiol Heart Circ Physiol 285:H2770–H2779

Zuk PA, Zhu M, Mizuno H, Huang J, Futrell JW, Katz AJ, Benhaim P, Lorenz HP, Hedrick MH (2001) Multilineage cells from human adipose tissue: implications for cell-based therapies. Tissue Eng 7:211–228

Rev Physiol Biochem Pharmacol (2006)
DOI 10.1007/112_0601

N. Toda · K. Ayajiki

Phylogenesis of constitutively formed nitric oxide in non-mammals

Published online: 12 September 2006

Abstract It is widely recognized that nitric oxide (NO) in mammalian tissues is produced from L-arginine via catalysis by NO synthase (NOS) isoforms such as neuronal NOS (nNOS) and endothelial NOS (eNOS) that are constitutively expressed mainly in the central and peripheral nervous system and vascular endothelial cells, respectively. This review concentrates only on these constitutive NOS (cNOS) isoforms while excluding information about iNOS, which is induced mainly in macrophages upon stimulation by cytokines and polysaccharides. The NO signaling pathway plays a crucial role in the functional regulation of mammalian tissues and organs. Evidence has also been accumulated for the role of NO in invertebrates and non-mammalian vertebrates. Expression of nNOS in the brain and peripheral nervous system is widely determined by staining with NADPH (reduced nicotinamide adenine dinucleotide phosphate) diaphorase or NOS immunoreactivity, and functional roles of NO formed by nNOS are evidenced in the early phylogenetic stages (invertebrates and fishes). On the other hand, the endothelium mainly produces vasodilating prostanoids rather than NO or does not liberate endothelium-derived relaxing factor (EDRF) (fishes), and the ability of endothelial cells to liberate NO is observed later in phylogenetic stages (amphibians). This review article summarizes various types of interesting information obtained from lower organisms (invertebrates, fishes, amphibians, reptiles, and birds) about the properties and distribution of nNOS and eNOS and also the roles of NO produced by the cNOS as an important intercellular signaling molecule.

N. Toda (✉)
Toyama Institute for Cardiovascular Pharmacology Research,
7-13, 1-Chome, Azuchi-machi,
Chuo-ku, Osaka, Japan
e-mail: n.toda.toyama-bldg@orion.ocn.ne.jp

N. Toda · K. Ayajiki
Shiga University of Medical Science, Department of Pharmacology,
Seta, Ohtsu, Japan

Abbreviations

7-NI	7-nitroindazole
ACE	angiotensin I-converting enzyme
ACh	acetylcholine
ADMA	asymmetric dimethylarginine
BH_4	tetrahydrobiopterin
CGRP	calcitonin gene-related polypeptide
cNOS	constitutive NOS
CO	carbon monoxide
DAF-2	diaminofluorescein-2
EDHF	endothelium-derived hyperpolarizing factor
EDRF	endothelium-derived relaxing factor
eNOS	endothelial NOS
EPP	endplate potential
FAD	flavin adenine dinucleotide
FMN	flavin mononucleotide
GABA	γ-aminobutyric acid
GMP	guanosine monophosphate
GAP-43	growth-associated protein-43
iNOS	inducible NOS
IP_3	inositol trisphosphate
KLF	Kruppel-like factor
L-NA	N^G-nitro-L-arginine
L-NAME	L-NA methylester
L-NIL	L-N^6-(1-iminoethyl)lysine
L-NIO	L-N^6-(1-iminoethyl)ornithine
L-NMMA	N^G-monomethyl-L-arginine
NADPH	reduced nicotinamide adenine dinucleotide phosphate
NMDA	*N*-methyl-D-aspartate
nNOS	neuronal NOS
NO	nitric oxide
NOS	NO synthase
ODQ	1H-[1,2,4]oxadiazolo[4,3-a]quinoxalin-1-one
PACAP	pituitary adenylate cyclase-activating polypeptide
PDE-5	phosphodiesterase type 5
SNP	sodium nitroprusside
VIP	vasoactive intestinal polypeptide

Introduction

The discovery of endothelium-derived relaxing factor (EDRF) by Furchgott and Zawadzki (1980) provided a clue leading to the discovery that nitric oxide (NO) is one of the important intercellular mediators for the regulation of biological functions in mammals (Furchgott 1988; Ignarro et al. 1987; Palmer et al. 1987). The NO synthesizing enzyme (NO synthase: NOS) was first found in rat cerebellum by Bredt and Snyder (1990), and Moncada et al. (1988) and Palmer et al. (1988a) determined how NO is synthesized from the substrate

L-arginine. Neuronal NOS (nNOS) and endothelial NOS (eNOS) are constitutively expressed, while NOS in macrophages is not constitutive but is induced by cytokines and polysaccharides (iNOS). NO produced via catalysis by these NOS isoforms physiologically controls specific functions of the cardiovascular, central and peripheral nervous, and immune systems or in contrast may impair these systems if NO is massively formed and liberated. From the reports to date, there is abundant information about NOS expression and NO actions on mammalian organs and tissues. In addition, accumulated evidence in recent years has revealed that NO is an important mediator in the entire phylogenetic scale, including invertebrates and non-mammalian vertebrates. It would be quite intriguing to look back in time, taking a historical view of the development of NO synthases and biological importance of NO formed by NOS in lower organisms. Wilken and Huchzermeyer (1999) found that enzymatic NO synthesis in the yeast *Candida tropicalis* resembles the one in animal tissues with respect to the substrate L-arginine as well as its sensitivity to enzyme inhibitors.

This review article describes the phylogenesis of nNOS and eNOS expressions and the functional roles of NO synthesized only by constitutive NOS (cNOS) in invertebrates, fishes, amphibians, reptiles, and birds, as well as including some information about the ontogenesis of cNOS in mammals for comparison.

Synthesis and actions of NO in mammals

NOS isoforms

NO-synthesizing enzyme was first isolated, purified, and cloned from the rat cerebellum by Bredt and Snyder (1990). The nNOS (or isoform I) constitutively expressed in the brain, peripheral nerves, and kidneys is mostly a soluble enzyme requiring Ca^{2+} and calmodulin in the presence of tetrahydrobiopterin (BH_4), heme, NADPH, flavin adenine dinucleotide (FAD), and flavin mononucleotide (FMN) for activation. Isoform II of NOS is not constitutively expressed, but bacterial lipopolysaccharide and cytokines induce it in macrophages (Stuehr et al. 1991a; Hevel et al. 1991), thus leading it to be called inducible NOS (iNOS). In the present article, the roles of NO formed by iNOS are not included. Isoform III or eNOS is constitutively expressed mostly in endothelial cells (Förstermann et al. 1991). BH_4, as a cofactor, plays quite an important role in producing NO by eNOS. An insufficiency of BH_4 leads to uncoupling of the L-arginine/NO pathway, resulting in reduced NO production and increased formation of superoxide anions by NOS (Kwon et al. 1989; Werner et al. 1995; Shinozaki et al. 1999). The biosynthesis of BH_4 is controlled by hormones, cytokines, and certain immune stimuli (Werner-Felmayer et al. 2002).

NO production

NO is produced when L-arginine is transformed via N^G-hydroxy-L-arginine (Stuehr et al. 1991b) to L-citrulline by catalysis of NOS in the presence of O_2 and cofactors (Fig. 1). It was reported that half-saturating L-arginine concentrations (K_m) were 1.4 μM (Bredt and Snyder 1990) and 2.2 μM (Schmidt et al. 1991) for nNOS and 2.9 μM for eNOS (Pollock et al. 1991). Availability of L-arginine for NO synthesis is regulated by de novo L-arginine production, L-arginine transport across the cell membrane, and L-arginine breakdown by arginase (Hallemeesch et al. 2002). L-Arginine is synthesized intracellularly from L-citrulline through argininosuccinate synthase and argininosuccinate lyase (Wiesinger 2001). L-Arginine is also supplied by uptake from the extracellular space through a cationic amino acid transporter system (system y+; Baydoun and Mann 1994; Closs et al. 1997). Arginases I and II, arginine-degrading enzymes, do not compete with NOS in the enzymatic reaction since the K_m values for arginases are 2 to 20 mM, values that are about 1,000 times higher than those for NOS (Grody et al. 1987; Griffith and Stuehr 1995). However, once arginases are induced under pathological conditions, arginases and NOS may compete for the utilization of L-arginine.

eNOS is present mainly in particulate fractions (Förstermann et al. 1991) and binds to caveolin-1 in the caveolae that are microdomains of the plasma membrane (Garcia-Cardena et al. 1996; Shaul et al. 1996). Caveolin-1 inhibits eNOS activity, and this interaction is regulated by Ca^{2+}/calmodulin (Ju et al. 1997; Michel et al. 1997). eNOS intracellularly migrates in response to increased cytosolic Ca^{2+} in the presence of calmodulin (Fig. 1) and is activated for NO production (Pollock et al. 1991). The transmembrane influx of Ca^{2+} and its mobilization from intracellular stores are caused by stimulation of drug receptors, such as muscarinic, peptidergic, purinergic, and serotonergic receptors, in the endothelial cell membrane or by mechanical stresses applied to the endothelium. Later, it was noted that shear stress, insulin, or bradykinin induced the phosphorylation of $Ser^{1177/1179}$ of eNOS through phosphatidylinositol-3 (PI_3) kinase and the downstream serine/threonine protein kinase Akt (protein kinase B) (Fig. 1), resulting in enhanced NO synthesis without increase in intracellular Ca^{2+} (Dimmeler et al. 1999; Fulton et al. 1999; Harris et al. 2001). Furthermore, bradykinin activation of eNOS in cultured endothelial cells involved deinhibition of the enzyme through dephosphorylation at Thr^{497} (Fleming et al. 2001; Harris et al. 2001).

The synthesis of NO by eNOS and nNOS is inhibited by L-arginine analogs, including N^G-monomethyl-L-arginine (L-NMMA; Palmer et al. 1988b), N^G-nitro-L-arginine (L-NA; Rees et al. 1990; Toda et al. 1990a), L-NA methylester (L-NAME; Rees et al. 1990), and asymmetric dimethylarginine (ADMA; Vallance et al. 1992). 7-Nitroindazole (7-NI) is one of the most promising nNOS inhibitors so far introduced (Moore et al. 1993).

NO actions

The biological actions of EDRF/NO have been widely investigated in mammals. Endothelial NO causes vasodilatation, decreased vascular resistance, lowered blood pressure, inhibition of platelet aggregation and adhesion, inhibition of leukocyte adhesion and transmigration, and reduced smooth muscle proliferation. Further, it acts to prevent atherosclerosis.

Nonadrenergic noncholinergic inhibitory responses are mediated through NO synthesized by nNOS (Fig. 1) that plays an important role as a neurotransmitter from peripheral efferent nerves in blood vessels (Toda and Okamura 1990, 2003), gastrointestinal tracts (Bult et al. 1990; Toda et al. 1990b; Toda and Herman 2005), urinary tracts (Andersson et al. 1991), corpora cavernosa (Ignarro et al. 1990; Burnett et al. 1992), and tracheal muscles (Tucker et al. 1990). In the CNS, NO derived from nNOS activation acts as a modulator or transmitter for the regulation of neuronal activities.

NO/cyclic GMP pathway

Various nitro compounds capable of forming NO, such as nitroglycerin, sodium nitroprusside, and sodium nitrite, or NO gas activate soluble guanylyl cyclase and produce cyclic guanosine monophosphate (GMP) from GTP (Katsuki et al. 1977). NO derived from the

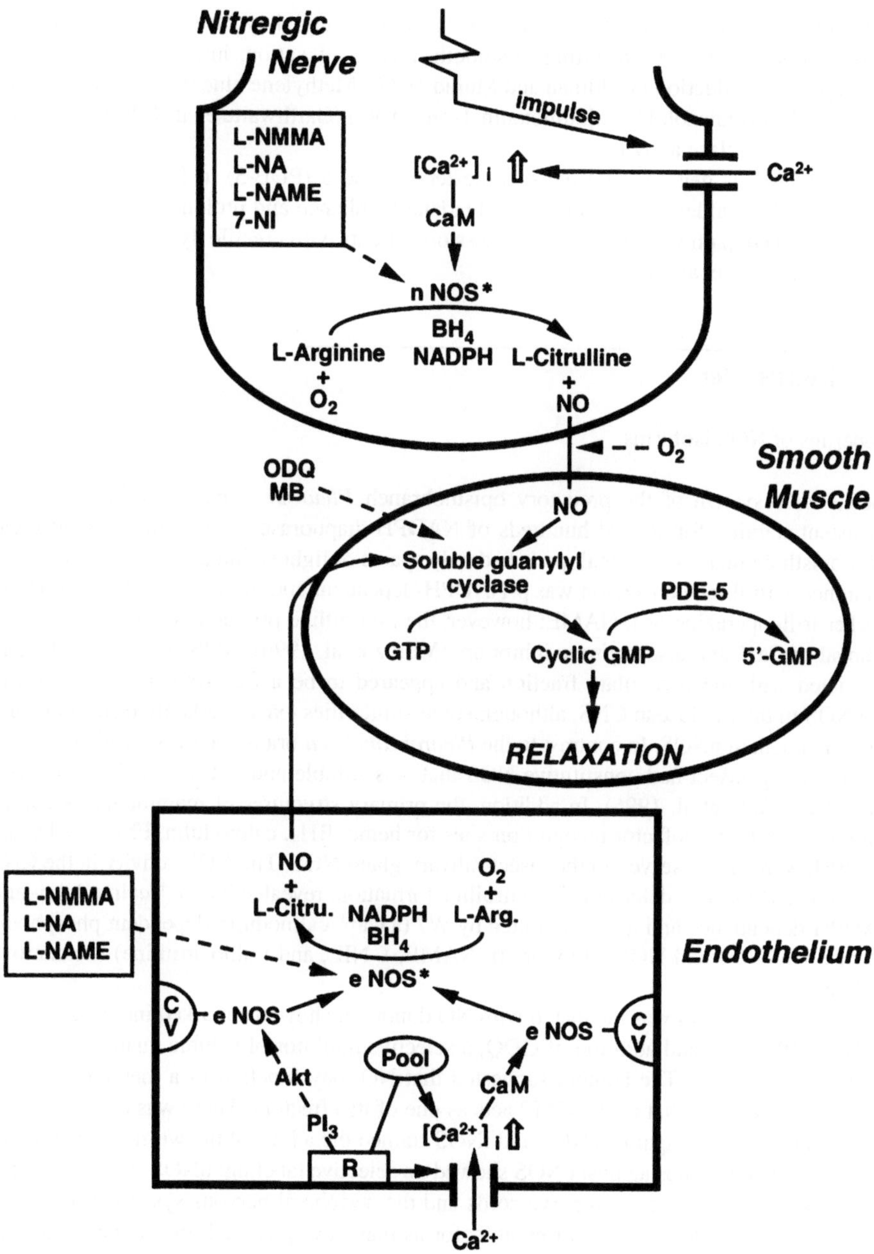

Fig. 1 Schematic presentation of actions of NO derived from the nitrergic nerve and endothelium on vascular smooth muscle. *CaM* calmodulin, *nNOS** activated nNOS, *eNOS** activated eNOS, O_2^- superoxide anion, *MB* methylene blue, *GTP* guanosine triphosphate, *5′-GMP* guanosine 5′-monophosphate, *PDE-5* phosphodiesterase-5, *CV* caveolin-1, *Pool* Ca^{2+} storage site, *R* receptive sites for drug and mechanical stimuli, *solid line* stimulation, *dotted line* inhibition. In nonvascular smooth muscles, such as those in gastrointestinal, urinary, and respiratory tracts, NO released from nitrergic nerves acts as an inhibitory neurotransmitter

endothelium or the nitrovasodilators activates soluble guanylyl cyclase and increases cyclic GMP production (Fig. 1), resulting in smooth muscle relaxation, intestinal secretion, and retinal phototransduction (Waldman and Murad 1987). Methylene blue (Gruetter et al. 1981) and 1H[1.2.4]oxadiazolo[4.3-a]quinoxalin-1-on (ODQ) (Garthwaite et al. 1985) inhibit soluble guanylyl cyclase activity.

Cyclic GMP is degraded by phosphodiesterase type 5 (PDE-5) to 5′-GMP. PDE-5 inhibitors, such as sildenafil, vardenafil, and tadalafil, enhance and prolong responses via the NO/cyclic GMP pathway; their effectiveness on patients with erectile dysfunction is widely recognized (Toda et al. 2005).

NO in invertebrates

Properties of NOS isoforms

The nervous system of the predatory opisthobranch *Pleurobranchaea californica* shows consistent staining for several hundreds of NADPH-diaphorase-positive neurons, whereas other opisthobranchs and cephalopods had only sparse or light staining for these neurons. L-Arginine/L-citrulline conversion was β-NADPH-dependent and inhibited by the calmodulin blocker trifluoperazine or L-NAME; however, the L-citrulline production was insensitive to removal of Ca^{2+} and also arginase inhibitors (Moroz et al. 1996). NOS activity was largely associated with the particulate fraction and appeared to be a Ca^{2+}-independent isoform. The NOS in the molluscan CNS, although some similarities exist, markedly differ in nature from mammalian nNOS. In contrast to the *Pleurobranchaea* brain, the salivary glands of the blood-sucking insect had constitutive NOS that was soluble and is Ca^{2+}- and calmodulin-dependent (Yuda et al. 1996). In addition, the primary structures of mammalian NOS, including the putative cofactor recognition sites for heme, BH_4, calmodulin, FMN, FAD, and NADPH, were all conserved in the insect salivary gland NOS. The NOS activity in the CNS of *Aplysia californica*, determined as citrulline formation, revealed its Ca^{2+}/calmodulin- and NADPH-dependence and it was inhibited by W7 (a Ca^{2+}/calmodulin-dependent phosphodiesterase inhibitor) and NOS inhibitors (L-NAME, L-NIL, and L-thiocitrulline) (Bodnarova et al. 2005).

When *Aplysia* CNS was incubated with NO donors, it showed a marked increase in basal cyclic GMP levels, and addition of ODQ, a selective inhibitor of soluble guanylyl cyclase, abolished this effect. The authors suggested that NO may function as a messenger in the molluscan CNS, and that cyclic GMP acts as one of its effectors. There was evidence indicating that antibodies against nNOS and iNOS stained the adults of the worms *Schistosoma mansoni*, and antibodies against eNOS showed no selective labeling; nNOS-like immunoreactivity was found in the main nerve cords and the peripheral nervous system (Kohn et al. 2001). They suggested that the different isoforms may have potential roles in neuronal signaling, reproduction, and development.

Localization of NOS and possible role of endogenous NO

There is supportive evidence for the idea that the NO signaling pathway is spread throughout the entire phylogenetic scale, including lower organisms from *Chordata to Mollusca* (Palumbo 2005).

Roles in the nervous system

Within minutes after crushing the nerve cord of the leech, the region of damage stained for NADPH-diaphorase and was immunoreactive for eNOS; on immunoblots of the leech CNS extract, the same antibody detected a band that was approximately the size of vertebrate eNOS (Shafer et al. 1998). Calmodulin was necessary for the eNOS immunoreactivity. Therefore, the authors concluded that an increase in NOS activity at lesions of the leech CNS appears to be among the earliest responses to injury and may be important for repairing axons. Similar beneficial actions of NO have been reported in injured tissues of the mammalian CNS within days of injury (Verge et al. 1992; Vizzard et al. 1995). Duan et al. (2005) provided evidence suggesting that in the leech, the first step in the repair process was the migration of microglia rapidly toward the axonal injury that was controlled in part by NO, which was generated immediately at the lesion and acted via a soluble guanylyl cyclase.

Bodnarova et al. (2005) provided evidence that NO may function as a messenger in the molluscan CNS and that cyclic GMP acts as one of its effectors. Pedal ganglia excised from the marine bivalve *Mytilus edulis* exhibited basal NO release, and the inhibition of basal NO release by L-NAME resulted in a great number of microglia in the incubation medium; this process involved two phases, a slow release during the early period and rapid egress occurring 18 h later (Stefano et al. 2004). The authors postulated that the spontaneous ganglionic NO release maintains/stabilizes microglia juxtaposed to neurons and that the excised ganglia at the various observation periods appear to reveal a transition of cNOS to iNOS derived NO. Kurylas et al. (2005) noted that NADPH-diaphorase histochemistry and NOS immunohistochemistry resulted in corresponding labeling patterns throughout the brain in the locust *Schistocerca gregaria*. Neuronal cell bodies in the locust midbrain were NADPH-diaphorase-positive, and highly prominent labeling occurred in the central complex, a brain area involved in sky-compass orientation, suggesting a crucial role for NO signaling in this brain area. Bicker (2005) has summarized how NO and cyclic nucleotides affect the wiring of nervous systems by regulating cell and growth-cone motility on the basis of the analysis of specific cell types in invertebrate models such as mollusks, insects, and the medicinal leech. In the embryos of the pond snail *Helisoma trivolvis*, NADPH-diaphorase histochemistry revealed NOS expression in serotonergic neurons and ciliary cells; NO donors and serotonin increased the rate of cilia-driven embryonic rotation within the egg capsule, and NOS inhibitors decreased the rotation rate (Cole et al. 2002). In cockroach pacemaker neurosecretory neurons, the membrane depolarization produced by neurohormone D, arachidonic acid, and eicosatetraynoic acid was related to the increase in membrane conductance for Ca^{2+}, Ba^{2+}, or Sr^{2+}, and this effect was abolished by L-NAME and ODQ, suggesting that any mechanism increasing the cyclic GMP level can induce noncapacitative Ca^{2+} entry (Wicher et al. 2004). Cayre et al. (2005) provided evidence that interneurons of the cricket mushroom body, the main integrative structures of the insect brain, synthesize NO and that the antennal nerve activity modulates progenitor cell proliferation and regulates NO production in brain structures where neurogenesis occurs in the adult insect. These authors suggested a key role for NO in environmentally induced neurogenesis.

The soluble guanylyl cyclase inhibitor LY-83583 blocked the stimulating effects of NO donors and serotonin, suggesting that NO has a constitutive cilioexcitatory effect in *Helisoma* embryos and that the effects of NO and serotonin work through a cyclic GMP pathway (Doran et al. 2003). NADPH-diaphorase activity was found in sensory epithelia and in the axial nerve cord of the arms of *Sepia officinalis*; injections of L-NAME produced an increase in the latency of prey paralysis but did not affect the flexibility of the manipulative behavior, suggesting that a direct role for NO in the motor skills of manipulative behavior is unlikely and that NO could play an important part in the transmission of chemical and tactile information (Halm et al. 2003). Robertson et al. (1994) noted that tactile learning tasks could be abolished by intramuscular injections of L-NAME in *Octopus vulgaris* and suggested that NO is required for tactile learning.

The terrestrial slug *Limax valentianus* could discriminate among the odor of l-octanol and l-hexanol; when L-NAME was injected shortly before the discrimination test, slugs could not discriminate between octanol and hexanol, whereas the retrieval of olfactory memory remained intact, indicating that the NO cascade may play a crucial role for the fine olfactory discrimination in *Limax* (Sakura et al. 2004). NADPH-diaphorase staining, anti-NOS immunocytochemistry, and the NO-indicator DAF-2 showed that cells throughout the optic anlage in the *Manduca sexta* contained NOS and produced NO; signaling via NO inhibited proliferation in the anlage (Champlin and Truman 2000). When exposed to low levels of ecdysteroid, NO production was stimulated and proliferation ceased, and when steroid levels were increased, NO production was decreased.

By using the method of direct single-cell microanalysis of intracellular concentrations of the major NOS-related metabolites in neurons from the pulmonate mollusk *Lymnaea stagnalis*, Moroz et al. (2005) noted that the esophageal motoneuron B2, but not neurons from the buccal, cerebral, and pedal ganglia, contained active NOS and exhibited an L-arginine/L-citrulline ratio susceptible to the NOS inhibitor. Intense NOS-immunoreactive neurons and important functional roles of nitrergic efferent nerves have widely been recognized in the mammalian esophagus (Toda and Herman 2005).

Other roles

Light production by fireflies was stimulated by NO in the presence of oxygen, and NO scavengers blocked bioluminescence induced by the neurotransmitter octopamine; NOS was expressed in the firefly lantern in cells interposed between nerve endings and the light-producing photocytes (Trimmer et al. 2001; Greenfield 2001). Trimmer et al. (2004) proposed that NO controls flashes by transiently inhibiting oxygen consumption in lantern mitochondria and by permitting direct oxidation of activated luciferin. Reversible inhibition by NO of the respiratory chain has been reported in mammalian mitochondria (Wink et al. 1993; Cleeter et al. 1994).

Immunostaining experiments showed that promastigotes from early passages of *Leishmania* in culture had a strong immunoreactivity against anti-cNOS and anti-eNOS, in comparison with the same parasite cultured for a long time, corroborating findings of a higher NO production by newly isolated parasites, following the growth curve (Genestra et al. 2003a). The presence of cNOS raised the possibility of a similar type of crosstalk or downregulation between the NO signaling systems in host cells and the lower eukaryotes such as *Leishmania sp.* Genestra et al. (2003b) also found evidence suggesting that the parasite NO pathway was important for establishment of the infection in murine macrophages.

Neuropeptides of the capa gene are typical of the abdominal neurosecretory system of insects (Predel et al. 2003). Pollock et al. (2004) showed the generality of capa peptide action—to stimulate NO/cyclic GMP signaling and increase fluid transport—across the *Diptera*, but not in the more primitive *Orthoptera*.

Pathological implications

In the gut of mice infected with the protozoan parasite *Trypanosoma cruzi*, Ca^{2+}-dependent nNOS/eNOS activity was decreased, whereas Ca^{2+}-independent eNOS activity was increased, suggesting that alterations in the NO system may be important in the pathogenesis of the gastrointestinal manifestations in Chagas' disease (Ny et al. 1999).

The fire ant *Solenopsis invicta* venom is composed of piperidine alkaloid components, which produced potent inhibition of NOS isoforms, rat cerebellar neuronal (nNOS), bovine recombinant endothelial (eNOS), and murine recombinant immunologic NOS (iNOS) (Yi et al. 2003). Isosolenopsin A (*cis*-2-methyl-6-undecylpiperidine), a naturally occurring fire ant piperidine alkaloid, inhibited the NOS isoforms; the IC_{50} (median inhibitory concentration) was approximately 18 μM for nNOS, 156 μM for eNOS, and greater than 1,000 μM for iNOS. The authors concluded that inhibition of NOS isoforms by isosolenopsin A and structurally similar compounds may have toxicological significance with respect to adverse reactions to fire ant stings. Although some of their toxicity is likely exerted by the iNOS expression, the present review article describes only toxic effects of various venoms in relation to NO formed by cNOS.

Weinberg et al. (2002) found that a semi-purified fraction of *Phoneutria nigriventer* spider venom induced endothelium-dependent relaxation of rat mesenteric arterial rings that was inhibited by L-NAME, but not by indomethacin, suggesting that this spider venom fraction participates in the liberation of NO from the arterial endothelium, in which a non-muscarinic mechanism might be involved in this effect, but mechanisms relating to prostanoids and bradykinin are not involved. Tx2-5, a sodium channel-selective toxin obtained from the venom of the spider *P. nigriventer*, produced penile erection by means of an NO mechanism; the effect of Tx2–5 was partially inhibited in mice pretreated with L-NAME and abolished in those treated with 7-NI (selective nNOS inhibitor) (Yonamine et al. 2004). 7-NI also abolished all the other symptoms of intoxication induced by Tx2-5, including salivation, respiratory distress, and death. They suggested that nNOS is the major player in the spider venom intoxication and that toxins from other animals may have as a major common target the activation of nNOS. The toxic fraction PhTx2 of the spider prolonged the inactivation and deactivation processes of sodium ion channels (Araujo et al. 1993). PhTx2 evoked ACh release from rat cortical synaptosomes and this effect was inhibited by the sodium channel blocker tetrodotoxin (Moura et al. 1998). NO released from the cavernous nerve appears to participate in the toxin-induced penile erection. There was evidence that neuronal release of NO was responsible for the increase in the intracavernous pressure associated with activation of central melanocortin receptors (Vemulapalli et al. 2001). Therefore, central actions of the toxin may also be considered. However, the hypothesis that hypersalivation in response to Tx2-5 is mediated by the increased release of NO is unlikely since NO donors injected into the medial septal area or medial preoptic nucleus attenuate, but L-NAME injections augment, salivary secretion induced by pilocarpine (Saad et al. 2002, 2003).

In rabbit corpus cavernosum strips, *Androctonus australis* and *Buthotus judaicus* scorpion venoms induced relaxations that were inhibited by L-NAME and 7-NI and also by tetrodotoxin, suggesting that NO released from nitrergic nerve fibers mediates the venom-induced cavernous muscle relaxation (Teixeira et al. 2001a). Similar findings with *Tityus serrulatus* scorpion venom were obtained in isolated human corpus cavernosum (Teixeira et al. 2001b). They suggested that the mechanism responsible for the actions of these venoms might be useful for understanding the development of priapism in cases of scorpion envenomation. An α-toxin isolated from *T. serrulatus* venom that was responsible for NO-mediated relaxation of the rabbit corpus cavernosum was identified as Ts3 after determination of Cys residues, N-terminal amino acid analysis, and proteolytic peptide mapping (Teixeira et al. 2003).

NOS in crustaceans

Biochemical assay on crayfish (*Procambarus clarkii*) eyestalk ganglia indicated that NOS activity was inhibited by L-NA, the calmodulin inhibitor fluphenazine-N-2-chloroethane, as well as by omitting β-NADPH or Ca^{2+} from the incubation medium, indicating that the eyestalk ganglia appear to have biochemical characteristics similar to mammalian cNOS (Lee et al. 2000). The localization of NOS-immunoreactivity suggests that NO may be involved in regulating visual processes and neuroendocrine function in crustaceans.

NO formed by constitutive NOS in fish

Vascular endothelium

Coronary arteries isolated from the rainbow trout contracted with ACh did not respond to histamine and bradykinin, and adenosine, ADP, ATP, or ACh caused only contractions (Small et al. 1990), these agents being recognized to elicit relaxation by releasing EDRF from the endothelium of mammalian blood vessels (Angus and Cocks 1989). NO donors, nitroglycerin and sodium nitroprusside (SNP), relaxed the trout coronary artery. Olson and Villa (1991) noted that ACh produced contractions in ring preparations from the trout aorta, efferent branchial and celiacomesenteric arteries, and anterior cardinal veins and that neither ACh nor bradykinin relaxed pre-contracted vessels; removal of the endothelium did not affect responses to these agonists. The Ca^{2+} ionophore A23187 produced an endothelium-dependent relaxation, which was abolished by the cyclooxygenase inhibitor indomethacin. SNP relaxed all vessels. These findings suggest that endothelium-derived prostanoids, but not NO, are involved in the trout vasculatures tested, and soluble guanylyl cyclase may participate in the response to exogenous NO.

ACh, SNP, and NO itself elicited contraction in the anterior mesenteric artery and posterior intestinal vein form the dogfish shark (Evans 2001). ACh constricted, but SNP or NO dilated, aortic rings from the eel, while SNP and NO constricted the rings from the hagfish (Evans and Harrie 2001). In isolated rings of the ventral aorta of the spiny dogfish shark, neither L-arginine nor SNP (nor NO itself) produced vasodilatation, whereas A23187 elicited endothelium-dependent dilatation that was not inhibited by L-NAME but was depressed by indomethacin, suggesting that prostanoids, most probably prostacyclin, are involved in the response (Evans and Gunderson 1998). Park et al. (2000) also concluded that A23187 appears to induce relaxation of the carp aorta through an endothelium- and cyclooxygenase-dependent mechanism and that this preparation does not produce NO as an EDRF nor does it respond to exogenously applied NO. Similar conclusions were drawn by Miller and Vanhoutte (2000) in the trout aorta and by Jennings et al. (2004) in the dorsal aorta and the intestinal vein of the Australian short-finned eel. Thus, Miller and Vanhoutte (2000) speculated that a phylogenetic appearance of an NO-sensitive mechanism for vasodilatation may be associated with the transition from water to air respiration. Bradykinin-induced cod celiac artery dilatation was abolished by indomethacin, and there was no evidence for the involvement of NO and leukotrienes (Shahbazi et al. 2001). In the dorsal aorta of the giant shovelnose ray, ACh caused constriction and SNP did not mediate any vasodilatation, suggesting that NO is not involved in the response to ACh (Donald et al. 2004). However, they noted that nicotine induced dilatation of the ring preparations that was not affected by L-NA or ODQ but was eliminated by disruption of the endothelium. In addition, indomethacin reduced the nicotine-induced relaxation, supporting the hypothesis that cyclooxygenase products are released from the endothelium.

On the other hand, evidence for the release of NO from the endothelium was demonstrated in the trout (*Oncorhynchus mykiss*) coronary vasculature, in which adenosine-induced vasodilatation was reduced by L-NA and also by chemical denudation of the endothelium (Mustafa and Agnisola 1998). Furthermore, Agnisola (2005) obtained evidence implicating a main role for NO as an amplifier of the adenosine-mediated coronary vasodilatation under hypoxia. The early larval arterial and venous vasculature of zebrafish in vivo responded to SNP with dilatation, and L-NAME caused a decrease in the vascular diameters; NOS immunoreactivity was demonstrated in endothelial cells of the dorsal vein (Fritsche et al. 2000). Basal release of NO may be involved in the vasodilatation. However, the authors did not elucidate the effect of EDRF-releasing agents, such as A23187, adenosine and ACh, nor the reversal by L-arginine of the L-NAME-induced vasoconstriction. NO may be derived partly from the endothelium but possible involvement of the gaseous molecule from perivascular nerves cannot be ruled out.

In unanesthetized trout, SNP reduced ventral aortic pressure, dorsal aortic pressure, and systemic resistance and increased cardiac output and heart rate, but did not affect central venous pressure, stroke volume, or gill resistance (Olson et al. 1997). In the heart of the eel *Anguilla anguilla*, NOS inhibitors, ODQ, and Triton X-100, a detergent that damages the endocardial endothelium, all increased stroke volume and stroke work; in contrast, L-arginine decreased these variables (Imbrogno et al. 2001). ACh at nanomolar concentrations induced a NO/cyclic GMP-dependent positive inotropism that required the integrity of the endocardial endothelium. The positive inotropism may be associated with an NO-mediated decrease in vascular resistance. Eddy and Tibbs (2003) noted that NOS inhibitors elicited tachycardia, possibly mediated by vasoconstriction through reduced NO synthesis in *Salmonid alevins*; this effect was reversed by L-arginine. Tota et al. (2005) obtained evidence emphasizing that NO is important in reshaping the angio-myoarchitecture of the teleost heart ventricle.

In the rainbow trout (Soderstrom et al. 1995) and crucian carp (Hylland and Nilsson 1995) in vivo, ACh induced an increase in cerebral blood flow velocity that was blocked by NOS inhibitors. The latter authors suggested that in the fish, NO-dependent blood flow-stimulating effects of ACh are mainly restricted to cerebral blood flow. In these studies, however, endothelium-dependency of the response was not determined; therefore, ACh may also release NO from vasodilator nerves. In the rainbow trout, oxytocin, but not arginine vasopressin, substance P, and bradykinin, increased cerebral blood flow velocity and L-NA suppressed the effect (Haraldsen et al. 2002). According to Nilsson and Soderstrom (1997), some other fishes appear to lack NO-mediated cerebral vasodilatation. In mammals, hypercapnia causes cerebral vasodilatation and increased cerebral blood flow through mechanisms that involve the production of NO (Iadecola 1992; Sandor et al. 1994). In contrast to air-breathing vertebrates, fishes depend much more on the blood O_2 tension than on the CO_2 tension for their respiratory regulation (Nilsson and Soderstrom 1997). Exposing the rainbow trout to hypercapnia resulted in an increase in cerebral blood flow velocity, whereas no change was observed in the crucian carp (Soderstrom and Nilsson 2000). None of the circulatory changes seen in the trout were blocked by L-NA, suggesting that the hypercapnia-induced increase in cerebral blood flow velocity is independent of NO production.

Table 1 Effects of EDRF-releasing agents (in mammals) and NO donors and EDRF identification in fish vasculatures

Authors	Species	Blood vessel	Stimulation	EDRF	Histochemistry	Comment
Small et al. 1990	Trout	Coronary artery	ACh, BK	None		NO donor, relaxation
			Adenosine	None		
Olson and Villa 1991	Trout	Aorta, arteries, veins	ACh, BK	None		
			A23187	PG		
Hylland and Nilsson 1995	Carp	Brain surface vasculature	ACh	NO[a]		
Mustafa and Agnisola 1998	Trout	Coronary vasculature	Adenosine	NO		
			L-Arginine	NO		
Evans and Gunderson 1998	Dogfish shark	Aorta	A23187	PG		NO donor, no relaxation
Park et al. 2000	Carp	Aorta	ACh, BK	None		NO donor, no relaxation
			A23187	PG		
Fritsche et al. 2000	Zebra fish (larva)	Artery and vein	L-NAME[b]	NO	NOS-IR (vein)	NO donor, relaxation NO only from endothelium?
Miller and Vanhoutte 2000	Trout	Aorta	A23187	PG		NO, no relaxation NO donor, relaxation
Evans and Harrie 2001	Eel, hagfish	Aorta	ACh	None		NO donor, relaxation, relax (eel) NO donor, contraction (hagfish) ACh, contraction
Evans 2001	Dogfish shark	Mesenteric artery	ACh	None		NO donor, contraction
Pellegrino et al. 2002	Eel	Branchial circulation	ACh	None		NO donor, contraction
Donald et al. 2004	Ray	Aorta, artery, vein	ACh	None		ACh, contraction
			Nicotine	PG		
Jennings et al. 2004	Eel	Aorta, intestinal vein	A23187	PG		NO donor, relaxation

[a] Involvement of the endothelium undetermined
[b] No agonist used; this inhibitor induced contraction

From the investigations so far (Table 1), ACh and bradykinin do not appear to liberate EDRF from the endothelium in preparations isolated from peripheral vasculatures in fish, except for brain surface vasculatures in vivo, while A23187 would act on the endothelium to liberate vasodilating prostanoids as an EDRF, as evidenced by a number of investigators. Endothelial cells appear to lack functioning of cholinergic receptors; however, cholinergic receptors present in smooth muscle cells are responsible for vasoconstriction. On the other hand, in the isolated perfused nonworking heart of the trout and in larval arterial and venous vasculatures of the anesthetized zebrafish, evidence for the release of NO from the endothelium was obtained by the administration of adenosine and L-arginine or by treatment with L-NAME. In order to determine whether the different nature of EDRFs in the fish blood vessels is due to large arteries and veins versus microcirculation, different fish species, or else, further analytical studies are required with careful reference to information accumulated to date.

Contrary to other investigators, Pellegrino et al. (2002) noted that ACh, L-arginine, and NO donors SIN-1 and SNP caused vasoconstriction in a branchial basket preparation of the eel. ODQ inhibited the effects of SIN-1 and SNP, and 8-bromo-cyclic GMP induced vasoconstriction. In their experiments, NOS inhibitors unexpectedly caused vasoconstriction. Adenosine-induced vasoconstriction in an isolated perfused head and a branchial basket preparation from the dogfish and eel was blocked by the NOS inhibitor and ODQ, suggesting that the response is mediated by an NO–cyclic GMP mechanism (Pellegrino et al. 2005).

Perivascular nitrergic innervation

Histochemical studies on NOS-immunoreactivity or NADPH-diaphorase staining showed the presence of NOS-containing neurons in blood vessels of various fishes (Hylland and Nilsson 1995; Esteban et al. 1998; Renshaw and Dyson 1999; Jimenez et al. 2001; Donald et al. 2004; Jennings et al. 2004; Table 2). However, functional studies concerning the role of neurogenic NO in producing vasodilatation are quite limited in the fish, despite the fact that the functional role of efferent nitrergic nerves has clearly been elucidated in mammals (Toda and Okamura 2003). Jennings et al. (2004) noted that the nicotine-induced dilatation of the dorsal aorta and intestinal vein of the Australian short-finned eel *A. australis* was endothelium-independent and was blocked by treatment with L-NA, the nNOS-selective inhibitor N^{ω}-propyl-L-arginine, and ODQ; indomethacin did not affect the response to nicotine. Nicotine is commonly used to stimulate nitrergic nerves in mammalian blood vessels, in which vasodilatation and NO release induced by nicotine are susceptible to ganglionic blocking agents or tetrodotoxin. If this is the case in fish preparations, it would be valid to conclude that nerve-derived NO contributes to the maintenance of vascular tone in this species.

Schwerte et al. (1997) noted that electrical vagal nerve stimulation increased swimbladder blood flow in the European eel *A. anguilla* and this effect was not blocked by α- and β-adrenoceptor antagonists and atropine; vagotomy resulted in a marked decrease in blood flow. Further studies by these authors (Schwerte et al. 1999) provided no conclusive evidence that vasoactive intestinal polypeptide (VIP), NO, or adenosine is responsible for the effect of vagal stimulation. In mammals, postganglionic parasympathetic nerves contain nitrergic and cholinergic neurons innervating cerebral and ocular arteries, and neurogenic NO is involved in vasodilatation (Toda and Okamura 2003). According to Hylland and Nilsson (1995) and Soderstrom et al. (1995), ACh and SNP increased cerebral blood flow veloc-

Table 2 Histochemical demonstration and functional role of NOS-positive neurons in fish vasculatures

Authors	Species	Blood vessel	Histochemistry	Stimulation	NO-dependent dilatation	Comment
Hylland and Nilsson 1995	Carp	Cerebral blood flow velocity	NOS, yes	ACh	Yes[a]	Nerve-derived NO?
Soderstrom et al. 1995	Rainbow trout	Cerebral blood flow velocity		ACh	Yes[a]	Nerve-derived NO?
Kagstrom and Holmgren 1997	Trout	Intestinal artery	VIP, yes NOS, no			VIPergic nerve
Esteban et al. 1998	Rainbow trout	Hepatic artery Portal vein	NOS, yes			Suggest: nitrergic nerve functioning
Renshaw and Dyson 1999	Shark	Brain microvasculature	NOS, yes			
Funakoshi et al. 1999	Puffer fish	Branchial artery	NOS, yes			Nitrergic nerve, sensory and baroreception
Jimenez et al. 2001	Trout	Renal artery Glomerular arteriole	NOS, yes			Suggest: nitrergic nerve functioning
Zaccone et al. 2003	Catfish	Branchial vasculature	NOS, yes			Nitrergic nerve, sensory
Donald et al. 2004	Ray	Aorta	NOS, yes	Nicotine	No	Nitrergic nerve, function unknown
Jennings et al. 2004	Eel	Dorsal aorta Intestinal vein	NOS, yes	Nicotine	Yes	Nitrergic nerve

[a] NO derived from the nerve or endothelium, undetermined

ity in the crucian carp and rainbow trout and the response to ACh was abolished by L-NA or L-NAME. Whether or not the effect of ACh was endothelium-dependent was not determined. Histochemical studies have demonstrated the presence of NOS-containing neurons in a variety of fish blood vessels as shown in Table 2, whereas evidence for the presence of NOS-positive endothelial cells has been reported only in early larval arteries and veins of zebrafish (Fritsche et al. 2000). There is evidence supporting the idea that ACh stimulates nitrergic nerves to induce dilatation in monkey ciliary arteries (Toda et al. 1998) and pial arterioles from eNOS knockout mice (Meng et al. 1998). These findings may indicate that ACh-induced vasodilatation in the fish brain is associated with stimulation of nitrergic nerves rather than endothelial cells.

Renshaw and Dyson (1999) found that intense NOS activity occurred in the microvasculature of epaulette shark brain following exposure to a severe hypoxia environment and suggested that enhanced NO production in response to hypoxia might cause vasodilatation, which would play a role in continued neuronal survival. Esteban et al. (1998) found nNOS-immunoreactivities in the hepatic artery, portal vein, and extralobular biliary duct and suggested that nitrergic intrinsic nerves play a role in controlling hepatic blood flow and hepatobiliary activity. Immunohistochemical studies using a polyclonal nNOS antibody showed that the nitrergic plexus reached the rainbow trout kidney along the vasculature, and large arteries and glomerular arterioles of the tubular trunks were innervated by nitrergic nerve fibers, suggesting a role of nitrergic neural structures in the control of particular renal functions and blood flow (Jimenez et al. 2001).

Funakoshi et al. (1999) provided evidence supporting the hypothesis that nitrergic sensory neurons in the glossopharyngeal and vagal ganglia project their peripheral processes through the branchia rami to a specific portion of the branchial arteries in the puffer fish *Takifugu niphobles*, and they might play a role in baroreception. These authors did not find nNOS immunoreactivity and NADPH-diaphorase activity in the parasympathetic postganglionic neurons of the branchial nerve in the puffer fish, whereas Gibbins et al. (1995) demonstrated NADPH-diaphorase activity in parasympathetic postganglionic nerves innervating the gill arches of the codfish *Gadus morhua.* In mammals, baroreceptors are located in the carotid sinus at the bifurcation of the carotid artery and in the aortic arch, which are homologous to the branchial arteries of the first and the second gill arches of fishes, respectively (Burleson and Milsom 1993). It has been shown that NADPH diaphorase activity and NOS-immunoreactivity are present in the baroreceptive sensory nerve fibers innervating the carotid sinus in the rat (Hohler et al. 1994). From the findings obtained thus far, Funakoshi et al. (1999) suggested that the nitrergic sensory nerves innervating the efferent filament artery in *Takifugu* function as baroreceptors that transmit signals to the medullary visceral sensory column. Zaccone et al. (2003) also found a nitrergic innervation in most arteries of the gill and air sac of the Indian catfish. Although the origin of the nerve fibers was not determined, they postulated that both NADPH-diaphorase activity and nNOS-immunoreactivity were present in the sensory system.

In small arteries from the proximal intestine region of the rainbow trout, VIP elicited relaxations that were not affected by L-NAME and the lipoxygenase inhibitor esculetin, but were partially inhibited by indomethacin (Kagstrom and Holmgren 1997). VIP immunoreactivity was found in varicose nerve fibers in these vessels, but NOS immunoreactivity could not be demonstrated. They suggested that VIP is an endogenous vasodilating neuropeptide and that NO is not involved in the VIP-induced relaxation in the trout arteries.

Nitrergic innervation in nonvascular tissues

Functional roles of the nitrergic nerve were demonstrated more clearly in the fish stomach than in fish vasculatures. Green and Campbell (1994) noted that stimulation of the vagus nerve to isolated stomach preparations from the rainbow trout (*Salmo gairdneri*) treated with atropine caused relaxations that were profoundly reduced by L-NA; this inhibition was reversed by L-arginine. Therefore, they concluded that the vagal inhibitory postganglionic neurons are nitrergic in this tissue. Histochemical studies demonstrated the presence of NOS-immunoreactive and NADPH-diaphorase-positive nerve cells and fibers in the myenteric ganglia and nerve trunks throughout the gastrointestinal tract of the rainbow trout (Li and Furness 1993) and the goldfish *Carassius auratus* (Bruning et al. 1996). The latter authors also found that the intestinal sphincter, opening of the pneumatic duct, and swimbladder were densely equipped with stained neurons and fibers in the goldfish.

In an in situ saline-perfused posterior cardinal vein preparation from *O. mykiss* chromaffin cells, the catecholamine secretion during electrical field stimulation was abolished by SNP and markedly stimulated by L-NAME and the nNOS selective inhibitor 7-NI (McNeill and Perry 2005). Because removal of the endothelium by saponin did not affect NO production during electrical stimulation, the authors suggested that nNOS is the key isoform modulating the catecholamine secretion from trout chromaffin cells. Inhibitory effects of NO on catecholamine release have also been reported in mammals. ACh administered intra-arterially into the adrenal gland in anesthetized dogs increased the catecholamine secretion, this effect being enhanced by L-NAME and inhibited by the NO donor (Nagayama et al. 1998). In canine perfused adrenal glands, the catecholamine release evoked by the nicotinic cholinergic agonist was increased by L-NMMA and 7-NI, whereas L-NMMA, but not 7-NI, stimulated the basal catecholamine release, suggesting that NO from both neuronal and endothelial sources modulates evoked catecholamines release (Barnes et al. 2001). The nNOS activity was expressed in cultured bovine chromaffin cells and NOS inhibitors increased the basal catecholamine secretion (Vicente et al. 2002). In the head kidney of juvenile rainbow trout, the histochemical reaction for nNOS revealed the presence of this enzyme in some nerve fibers and ganglion cells and only rarely in chromaffin cells, suggesting an involvement of nerve-derived NO in the regulation of adrenal function (Gallo and Civinini 2001).

In the rainbow trout kidney, the collecting tubes, collecting ducts, large arteries, and glomerular arterioles of the tubular trunks were innervated by nNOS-immunoreactive nerve fibers, suggesting that nitrergic neural structures may be involved in the control of particular renal functions (Jimenez et al. 2001). The gas bladder of *Hoplerythrinus unitaeniatus* (air-breathing teleost) and *O. mykiss* (non-air breathing teleost) and the lung of *Lepidosiren paradoxa* (air-breathing dipnoan) all exhibited elevated cyclic GMP levels in response to SNP; only the *H. unitaeniatus* gas bladder responded to ACh with increased cyclic GMP, and L-NAME inhibited the effect of ACh (Staples et al. 1995). Thus, only the gas bladder of *H. unitaeniatus* appears to synthesize NO through cNOS. Ebbesson et al. (2005) showed that double-labeling for NOS immunoreactivity and Na^+,K^+-ATPase α-subunit messenger RNA (mRNA) revealed colocalization of Na^+,K^+-ATPase α-subunit mRNA and nNOS protein in putative cells of the gill of the Atlantic salmon and that NO donors inhibited the gill Na^+,K^+-ATPase activity, suggesting that the effect of NO on gill functions is mediated via an intracellular inhibition of Na^+,K^+-ATPase activity in putative chloride cells.

Role of NO in the central nervous system

The distribution of NOS-containing neuronal cell bodies, fibers, and putative nerve terminals was histochemically demonstrated in the diencephalons of the rainbow trout *O. mykiss* (Schober et al. 1993) and in the brain of the Atlantic salmon (Holmqvist et al. 1994). Development of NADPH-diaphorase activity from 20 h after fertilization to 5.5 days was demonstrated in the CNS of the cichlid fish, *Tilapia mariae* (Villani 1999). In the sunfish brain, NADPH-diaphorase histochemistry stained tanycytes almost exclusively (Ma 1993).

Oyan et al. (2000) found that cloning and sequencing of NOS isoforms in the cerebellum and optic tectum of the Atlantic salmon (*Salmo salar*) revealed a partial gene sequence of 560 bp corresponding to mammalian nNOS from cerebellum complementary DNA (cDNA), and the predicted protein sequence of identified salmon nNOS possessed 85% identity to that of mammalian nNOS (rat nNOS, Bredt et al. 1991; human nNOS, Nakane et al. 1993), suggesting that the arising of different vertebrate NOS isoforms is an evolutionarily old event. NOS immunoreactivity and NADPH-diaphorase activity partly coincided with the nNOS mRNA expression in the brain of the adult zebrafish but was present also in additional neuronal and non-neuronal cell types, indicating the occurrence of different NOS isoforms in the fish brain (Holmqvist et al. 2000). The nNOS may participate in neurotransmission or neuromodulation and in mechanisms related to the growth and neuronal plasticity.

Cioni et al. (1998) noted that an NADPH/Ca^{2+}-dependent NOS activity is present in the soluble and in the particulate fractions of teleost caudal spinal cord homogenates, both activities being inhibited by calmodulin inhibitors and by L-NAME. Western blot analysis using either anti-nNOS or anti-eNOS antibodies showed that the soluble enzyme corresponded to nNOS of mammals, whereas the particulate one was likely attributable to nNOS or eNOS (or both). In comparison with mRNAs encoding α- and β-subunits of soluble guanylyl cyclase, nNOS-immunoreactive and NADPH-diaphorase-positive neurons were more widely distributed in many cerebral neurons in the rainbow trout; however, wide overlaps of guanylyl cyclase subunit mRNAs and nNOS distribution were observed, suggesting a role for soluble guanylyl cyclase in various neuronal functions possibly via NO/cyclic GMP signaling in the fish brain (Ando et al. 2004).

Cyclic GMP immunoreactivity was determined in pineal photoreceptor cells, whereas NADPH-diaphorase-positive structures were located adjacent to these photoreceptor cells, suggesting a role for NO in pineal function in, for example, cyclic GMP-related events in the phototransduction process and the light-dark control of melatonin synthesis (Zipfel et al. 1999). Cioni et al. (2002) provided evidences that caudal neurosecretory cells of teleosts express nNOS and that NO acts as a putative modulator of the release of urotensins from the neurosecretory axon terminals. The nNOS was colocalized with arginine-vasopressin in a subpopulation of neurosecretory neurons of the teleost *Oreochromis niloticus*, suggesting that NO is implicated in the modulation of hormone release (Bordieri et al. 2003). NOS-like immunoreactivity was detected in ventral light organs of the mesopelagic fish *Argyropelecus hemigymnus*, and evidence for an important role of NO in the control of light emission was provided (Kronstrom et al. 2005). The NO effect seems unlikely to involve cyclic GMP. The nNOS-immunoreactive and NADPH-diaphorase-positive neurons were demonstrated in Muller cells, the radial glia of the retina, of salamander, goldfish, and catfish (Liepe et al. 1994).

Fish venom

Tetrodotoxin is a potent neurotoxin found in a variety of creatures, including the puffer fish *Fugu*, which is considered a delicacy in Japan. Tetrodotoxin blocks the rising phase of nerve action potentials via blockade of sodium conductance (Narahashi et al. 1964; Hagiwara and Nakajima 1966; Kao 1972). Therefore, in association with skeletal muscle paralysis due to blockade of motor nerve action potentials, tetrodotoxin inhibits responses to NO released by nitrergic nerve action potentials such as vasodilatation (Toda and Okamura 2003), gastrointestinal smooth muscle relaxation (Toda and Herman 2005), urinary tract sphincter relaxation (Andersson and Alm 2000), and penile erection (Toda et al. 2005) in mammals.

Stonustoxin isolated from the venom of the stonefish *Synanceja horrida* induced endothelium-dependent relaxations of the rat aorta that were inhibited by L-NAME, hemoglobin, and methylene blue, suggesting that the NO–cyclic GMP pathway is involved in the relaxation (Low et al. 1993; Sung et al. 2002). The venom of the soldierfish *Gymnapistes marmoratus* also produced relaxations in rat isolated endothelium-intact aortae, but no response was induced in the endothelium-denuded aorta; the venom-induced relaxation was inhibited by an NOS inhibitor but was unaffected by atropine (Hopkins and Hodgson 1998). The lionfish (*Pterois volitans*) venom elicited the endothelium-dependent relaxation in porcine coronary arteries that was potentiated by atropine but attenuated by an NOS inhibitor (Church and Hodgson 2002a). Church and Hodgson (2002b) have reviewed the cardiovascular effects of fish venoms in relation to the release of NO from the endothelium, smooth muscle contraction, and atrial contractility.

NO formed by constitutive NOS in amphibians

Cardiovascular system

In the working heart of the frog *Rana esculenta*, which lacks coronary vasculature, *en face* confocal scanning laser microscopy demonstrated the presence of NOS as a cytoplasmic constituent of the endocardial endothelial cells; L-NAME and methylene blue increased stroke volume and stroke work, whereas SNP decreased these parameters (Sys et al. 1997). Therefore, the endocardial endothelium appears to produce amounts of NO sufficient to modulate ventricular performance.

In the isolated aortic arches of the leopard frog *R. pipiens*, ACh and SNP caused relaxations; damage of the endothelium abolished the response to ACh but had no effect on the relaxation to SNP (Knight and Burnstock 1996). L-NAME inhibited the ACh-induced vasodilatation, suggesting that NO formed in the endothelium appears to mediate the relaxation to ACh. On the other hand, Broughton and Donald (2002) noted that there were no NADPH-diaphorase-positive and NOS-immunoreactive cells in the endothelium of central aortae of the toad *Bufo marinus*, whereas nNOS immunoreactivity was present in the perivascular nerves of the aorta. Relaxation induced by ACh was not dependent on the presence of the endothelium but was reduced by the nNOS inhibitor vinyl-L-NIO or abolished by L-NA, ODQ, and atropine. The authors proposed that an endothelial NO system is absent in the toad aorta and that NO generated by nNOS from perivascular nerves mediates vasodilatation. Activation of muscarinic receptors in nerve terminals appears to be involved. Similar results on neuronal muscarinic receptors were also obtained in the rat small intestine (Olgart and Iversen 1999). Broughton and Donald (2005) also provided no evidence for eNOS in the ventral abdominal vein and vena cava of the toad. They found that ACh-induced relaxation was endothelium-independent and was blocked by the nNOS-inhibitor in phenoxybenzamine-treated preparations, suggesting that nitrergic nerves rather than an endothelial NO system are involved in NO-mediated vasodilatation. Based on histochemical studies on the gastrointestinal tract of the axolotl *Ambystoma mexicanum*, Badawy and Reinecke (2003) found that the VIP/NOS-immunoreactive fibers were associated with submucosal blood vessels; thus, they suggested that these fibers may be involved in the regulation of submucosal blood flow. Injections of L-NAME into the sciatic artery elicited an increase in vascular resistance and a decrease in blood flow in the sciatic artery in anesthetized toads, whereas systemic blood pressure was unaffected by the NOS inhibitor (Rea and Parsons 2001).

Central nervous system

The NO–cyclic GMP pathway was immunohistochemically detected in the brain and pituitary of the aquatic toad *Xenopus laevis* (Allaerts et al. 1998). The nNOS and iNOS were potential sources of endogenous NO production in cultured cells of the pars intermedia of *Xenopus* (Allaerts et al. 2000). The authors provided evidence showing that endogenous NO production may be physiologically relevant under conditions where protection against oxidative damage is needed in non-mammalian vertebrates. Histochemical studies revealed that nNOS was expressed in the intermediate lobe of the frog pituitary and in cultured pituitary melanotrophs; whole cell patch-clamp recordings showed L-arginine and SNP provoked an inhibition of the current evoked by γ-aminobutyric acid (GABA) and L-NA produced a delayed increase of the GABA-evoked current amplitude (Castel et al. 2000). In addition, formation of cyclic GMP was enhanced by L-arginine and inhibited by the Ca^{2+}/calmodulin complex blocker W7. Thus, they concluded that NO, produced by Ca^{2+}/calmodulin-dependent NOS, appears to exert an autocrine inhibitory effect on the GABA-evoked current in frog melanotrophs.

In an in vitro brainstem preparation from North American bullfrog *R. catesbeiana*, addition of SNP or L-arginine caused increases in respiratory-related burst frequency, and L-NA reduced the burst frequency or abolished the neural activity, suggesting that production of NO, probably via nNOS, provides an excitatory input to the respiratory central pattern generator in the amphibian brainstem (Hedrick and Morales 1999). In isolated brainstem preparations from pre-metamorphic tadpoles of the bullfrog, L-NA increased fictive gill ventilation and lung burst frequency; in contrast, L-NA applied to the post-metamorphic tadpole brainstem decreased lung burst frequency (Hedrick et al. 2005). This changing role for NO coincides with the shift in importance in the different respiratory modes during development in amphibians; that is, pre-metamorphic tadpoles rely on gill ventilation, whereas post-metamorphic tadpoles have lost the gills and are obligate air-breathers using lungs for gas exchange. Heinrich et al. (2003) demonstrated with electron microscopy that eNOS and nNOS immunoreactivity colocalized in the same cell types in the saccule maculae of the frog *R. pipiens*, whereas these cNOS isoforms separately localized in other cells of the inner ear.

The earliest NOS-immunoreactive neurons were observed in the inferior reticular nucleus in the caudal rhombencephalon at embryonic stage 30 of the urodele amphibian *Pleurodeles waltl*, suggesting that the appearance of nitrergic cells reveals the first involvement of this system in reticulospinal control, likely influencing locomotor behavior (Moreno et al. 2002a, b). Exogenous NO, supplied to the *Xenopus* tadpole brain in vivo, decreased the number of proliferating cells and the total number of cells in the optic tectum, and conversely, inhibition of NOS activity increased the number of proliferating cells, suggesting that NO is an essential negative regulator of neuronal precursor proliferation during brain development (Peunova et al. 2001).

Using dorsal root ganglia and the dorsal horn of the developing and adult frog *R. esculenta*, Cristino et al. (2004) provided evidence that in limb dorsal root ganglia neurons, NOS parallels cell differentiation and limb development during metamorphosis and that NO production plays a key role in the maturation of sensory functions that subserves in amphibians the transition from swimming to tetrapod locomotion.

Other organs and tissues

On the basis of histochemical studies on the gastrointestinal tract of the urodele *A. mexicanum*, Maake et al. (2001) postulated that nitrergic nerves, together with those containing serotonin, VIP, substance P, and gastrin/cholecystokinin immunoreactivity, likely set up the fine regulation of gastrointestinal motility and blood flow.

L-NMMA, but not D-NMMA, caused a decrease in hydraulic conductivity of frog mesenteric capillaries, and L-arginine prevented the effect of L-NMMA, supporting the authors' hypothesis that inhibition of NO synthesis decreases capillary hydraulic conductivity (Rumbaut et al. 1995). Rea and Parsons (2001) reported that injections of L-NAME into empty-bladder toads produced a decrease in water uptake across the pelvic patch in association with a decrease in blood flow in the sciatic artery; thus, they hypothesized that during cutaneous drinking, blood flow into the capillary bed of the pelvic patch is regulated by NO. On the other hand, SNP attenuated the increase of osmotic water flow elicited by arginine–vasotocin and elevated cyclic GMP production in isolated urinary bladders of the frog *R. temporaria* treated with zaprinast, an inhibitor of the cyclic GMP-specific PDE-5; a NOS immunohistochemical study revealed a positive staining for nNOS in the mucosal epithelium, suggesting that endogenous NO may be involved in regulation of water osmotic permeability (Fock et al. 2004).

Kerschbaum et al. (2000) provided evidence suggesting that kidneys of the clawed frog *X. laevis* contain at least nNOS, but may contain an additional NOS isoform, which is less sensitive to calmodulin. It was suggested that NO is an intra-renal paracrine and/or autocrine factor, which may modulate the adaptations of frog renal functions to seasonal changes through the action of the cyclic GMP generated by guanylyl cyclase activation (Fenoglio et al. 2004). NOS was exclusively expressed in chromaffin cells of the frog adrenal gland, and SNP inhibited the stimulatory effects of ACTH, angiotensin II, and endothelin-1 on corticosterone and aldosterone secretion, providing evidence for a modulatory role of NO on adrenocortical cell activity (Cartier et al. 2001).

The NADPH-diaphorase histochemical technique revealed the presence of NOS in cell bodies and presumed processes of perisynaptic Schwann cells at the frog neuromuscular junction (Descarries et al. 1998). NO donors decreased endplate potential (EPP) amplitude as well as the frequency of miniature EPPs, while the NO scavenger hemoglobin, the nNOS inhibitor 3-bromo-7-nitroindazole, and the soluble guanylyl cyclase inhibitor LY-83583 increased the EPP amplitude at the frog neuromuscular junction, indicating that NO appears to regulate transmitter release via a cyclic GMP-dependent mechanism (Thomas and Robitaille 2001). Brunelli et al. (2005) provided evidence for an involvement of the NO system in the epidermis of *Triturus italicus* (Amphibia, Urodela) during the functional maturation phase starting at the climax and preceding structural rearrangements during metamorphosis, emphasizing the putative functional importance of eNOS and iNOS expressions related to a distinct phase of the biological cycle.

NO formed by constitutive NOS in reptiles

Cardiovascular system

Axelsson et al. (2001) found that the administration of L-NAME increased the basal tone of ring preparations of the aortic anastomosis in the estuarine crocodile, and they postulated a tonic release of NO in the preparation. In addition, immunohistochemistry revealed the presence of NOS- and tyrosine hydroxylase-containing nerve fibers in the adventitia and adventitio-medial border of the aortic anastomosis, suggesting that the crocodile vascular tone is reciprocally controlled by nitrergic and adrenergic mechanisms. Donald and Broughton (2005) noted that arteries and veins contained a plexus of NOS-positive perivascular nerves in reptiles, amphibians, and fishes, whereas NADPH-diaphorase-positive and eNOS-immunoreactive cells were seen only in the endothelium of reptiles.

In the anesthetized turtle *Trachemys scripta*, SNP induced systemic vasodilatation with no change in pulmonary vascular resistance, and L-NAME increased systemic vascular resistance while pulmonary resistance was unchanged, suggesting that NO has an important role in maintaining systemic vascular tone but pulmonary vasculature does not seem to respond to NO (Crossley et al. 2000). Treatment with L-NAME caused an increase in systemic blood pressure as well as a decrease in systemic vascular conductance, but had no effect on pulmonary arterial blood pressure and pulmonary conductance in anesthetized ball python (Skovgaard et al. 2005a). Galli et al. (2005) found that NO caused a systemic vasodilatation resulting in a reduction in systemic resistance, but did not affect pulmonary resistance in the South American rattlesnake. NO was established to have a potential role in regulating vascular tone in the systemic circulation. NO may be important for maintaining basal systemic vascular tone in varanid lizards, pythons, and turtles through a continuous release of NO (Skovgaard et al. 2005b). In the reptile *Crocodylus porous*, NO and prostaglandins did not control the characteristic heart rate hysteresis response to heat, although NO was important in buffering blood pressure against changes in heart rate during cooling (Seebacher and Franklin 2004). So far obtained, the origin of NO from the endothelium, perivascular nerve, or both has not been elucidated.

In the freshwater turtle *T. scripta*, topical application of ACh induced an increase in cerebral blood flow velocity that was blocked by L-NAME, suggesting that NO is an endogenous vasodilator in the turtle brain (Hylland et al. 1996; Nilsson and Soderstrom 1997). Again, whether NO was derived from the endothelium or the perivascular nerve was not determined. Soderstrom et al. (1997) found that injections of L-NA caused a temporary stop in cerebral blood flow as well as a persistent increase in systemic blood pressure in the turtle, suggesting that an NO-dependent vasodilatory tonus affects both cerebral and systemic blood circulation. Their study also showed that the cerebrovascular dilatation to hypercapnia was NO-independent. A hypoxia-induced increase in cerebral blood flow velocity was not influenced by L-NA and aminophylline in the estuarine crocodile, indicating an NO- and adenosine-independent cerebral vasodilatation by hypoxia (Soderstrom et al. 1999).

Gastrointestinal tracts

In the estuarine crocodile *C. porous*, NOS-immunoreactive nerve cell bodies and fibers were found in all regions of the gut examined, suggesting that neurally released NO is likely to be involved in the control of gastrointestinal motility in the crocodile (Olsson and Gibbins 1999) as in mammals (Toda and Herman 2005). NOS-positive cells contained immunoreactivity to VIP in the crocodile gut (Olsson and Gibbins 1999). NADPH-diaphorase- and NOS-positive nerve cell bodies and fibers were also found in the gastrointestinal tract of lizard and snake (Lamanna et al. 1999). The nitrergic neurons containing galanin appeared to be more numerous than those containing VIP. It was hypothesized that nitrergic neurons are inhibitory motor neurons innervating the circular smooth muscle, and the colocalization of NOS and VIP in neurons enhances their inhibitory action. Innervation of NADPH-diaphorase-positive neurons, together with VIP- and substance P-immunoreactive neurons, were demonstrated in the intestine of the lizard *Podarcis hispanica* (Martinez-Ciriano et al. 2000). In the gut of reptiles that had been fasted and those that had ingested a large meal, there were no major changes in the innervation by nerves containing NOS, calcitonin gene-related peptide (CGRP), galanin, pituitary adenylate cyclase-activating polypeptide, somatostatin, substance P/neurokinin, or VIP-like immunoreactivity (Holmberg et al. 2002).

Electrical field stimulation evoked relaxations in the presence of guanethidine and indomethacin in circular muscle strips of the Agama lizard esophagus, and the neurally induced relaxation was abolished by L-NAME, but not D-NAME, and L-arginine prevented the inhibitory effect of the NOS inhibitor (Knight and Burnstock 1999). SNP relaxed the strips, but ATP and VIP had no effect. The neurogenic inhibitory response of the lizard esophagus muscle appears to involve the L-arginine/NOS pathway. The CGRP-induced relaxation in *Iguana iguana* gallbladder strips was blocked by L-NAME, suggesting that this relaxation is mediated at least in part by NO release from nerves stimulated by CGRP (Kline and Pang 1994).

Central and peripheral nervous systems

Bruning et al. (1994a) histochemically determined colocalization of NADPH-diaphorase and NOS in the brain and spinal cord of the turtle *Pseudemys scripta elegans* and suggested that NO acts as a messenger molecule in different areas of the reptilian brain and spinal cord; in certain areas, the pattern of expression of NOS appears to have evolved before the radiation of the present mammalian, avian, and reptilian species. In the trigeminal ganglia of the infrared-sensitive crotaline snake *Trimeresurus flavoviridis*, nNOS-positive neurons were revealed; however, the positive neurons were more abundant in small and large neurons than in medium-sized neurons, which include most of the infrared-sensitive neurons, suggesting that nNOS may be involved in physiological functions such as the transmission of tactile, vibrotactile, and nociceptive sensations, rather than in infrared sensory processing (Moon et al. 2002). On the other hand, eNOS in the trigeminal ganglia of crotaline snakes appears to be involved in constitutive neurotransmission in these ganglia (Moon et al. 2004). In the female of the whiptail lizard species, which displayed copulatory behaviors indistinguishable from males when gonadectomized and treated with testosterone, treatment with a NOS inhibitor eliminated male-like behavior; the deficit was principally in mounting, suggesting that sexual motivational systems were affected, rather than consummatory mechanisms (Sanderson et al. 2005).

Seven days after transection of the sciatic nerve the NADPH-diaphorase activity increased in neurons of the dorsal root ganglia of the turtle *T. dorbigni*; a similar increase was also demonstrable in spinal glia and endothelial cells, suggesting the role of NO in hyperalgesia and neuronal regeneration or degeneration (Partata et al. 1999). In the first month post-caudotomy, co-induction of NOS, growth-associated protein-43 (GAP-43), and cell death repressor Bcl-2 were observed in spinal motoneurons that innervated the lizard regenerated tail, suggesting that the co-induction may be part of the molecular repertoire of injured motoneurons committed to survival and axon regeneration, and favor a role of NOS in motoneuron plasticity (Cristino et al. 2000a, b).

Graves et al. (2004) found that nNOS was present at the lizard neuromuscular junction and the synthesis and extracellular diffusion of NO was necessary to evoke both the immediate depression and delayed enhancement of ACh release by muscarine. They suggested that NO seems to be produced in either the surrounding perisynaptic Schwann cells or muscle.

NADPH-diaphorase-containing neurons were found throughout the lizard *Podarcis s. sicula* oviduct (Lamanna et al. 2001). The positive neurons were primarily located in the reproductive oviduct, and the vagina revealed a reactive nerve population denser than elsewhere; the positive neuron densities and the 17β-estradiol plasma levels coincided throughout the sexual cycle. The authors supposed that NO neurons play an estrogen-dependent role in the lizard oviduct muscle motility.

Retina

The intracellular application of L-arginine or cyclic GMP blocked gap junctions between horizontal cells of the turtle retina, suggesting that the L-arginine/NO/cyclic GMP pathway is present in retinal horizontal cells (Miyachi et al. 1990). The immunohistochemical localization of NOS in photoreceptors, amacrine cells, putative ganglion cells (Blute et al. 1997), and bipolar and horizontal cells (Haverkamp and Eldred 1998a) indicates that NO may function at several levels of visual processing in the turtle retina. Haverkamp and Eldred (1998b) demonstrated the localization of NADPH-diaphorase-positive retinal efferent cell bodies in the turtle brain, and they suggested the modulation of retinal function by NO. Confocal light microscopy showed eNOS-immunoreactive labeling in Muller cells of the retina from mammalian and non-mammalian vertebrates, including the turtle, and in axon terminals of turtle and fish horizontal cells (Haverkamp et al. 1999). The ultrastructural localization of nNOS-like immunoreactivity was detected in all layers of the turtle retina (Cao and Eldred 2001). According to these authors, the diverse subcellular localization of nNOS indicates that NO may play distinct functional roles in many retinal cells.

Haverkamp et al. (2000) noted that GABA and glycine colocalized with NADPH-diaphorase in amacrine cells of the turtle retina, suggesting that NO can modulate the release of GABA and glycine in the retina. Cyclic GMP-like immunoreactivity was increased in response to GABA receptor antagonists in the retina; the stimulating effect was blocked by the NOS inhibitor or glutamine-receptor antagonists, suggesting that inhibitory synaptic pathways involving GABA receptors work through excitatory glutamine receptors to regulate the NO/cyclic GMP signal transduction pathway (Yu and Eldred 2003). The same authors (2005) further provided evidence that NO modulates the levels of the inhibitory transmitter GABA and glycine through specific biochemical mechanisms in different retinal cell types. The evidence of nNOS-immunoreactivity in H2 and H3 horizontal cells of the turtle retina, together with available physiological evidence, suggests that NO may be involved in electrical coupling and/or modulation of synaptic input to these types of cells (Cuenca et al. 2000). The turtle retina stimulated by NO donors (Blute et al. 1998) and *N*-methyl-D-aspartate (NMDA) or kainic acid (Blute et al. 1999) showed increases in cyclic GMP-like immunoreactivity in amacrine cells. The NO release in response to NMDA from the turtle retina was visualized by diaminofluorescein-2 (DAF-2); the NO signal was primarily confined to within 10 mm of the source, which suggests that NO may not diffuse freely through the retina (Blute et al. 2000). Such limited spread was not predicted and suggests that NO signal transduction may be more selective than suggested in mammals. Blute et al. (2003) also found that nicotinic ACh receptors were immunohistochemically localized in the turtle retina and stimulation with nicotine increased NO production; select populations of amacrine cells responded to NO with increased levels of cyclic GMP-like immunoreactivity, suggesting that ACh can activate the NO/cyclic GMP signal transduction pathways. Carbon monoxide (CO) stimulated increases in cyclic GMP in bipolar and amacrine cells of the turtle retina, and the increases were blocked by L-NAME, leading Cao and Eldred to postulate that the response to CO may be mediated by CO releasing NO de novo or displacing NO from its intracellular storage pools, if any is present (Cao and Eldred 2003).

Levy et al. (2004) provided evidence supporting the hypothesis that background illumination triggers NO synthesis in the distal turtle retina and that NO acts to adjust the modes of visual information processing in the outer plexiform layer to the conditions required during continuous background illumination. Savchenko et al. (1997) found that *S*-nitrocysteine, an NO donor, activated cyclic nucleotide-gated channels in isolated lizard cones and that the NO donor or a membrane-permeant analog of cyclic GMP triggered the release of neurotransmitter from the cone terminals; the NO-induced transmitter release was suppressed by guanylyl cyclase inhibitors. It was suggested that these findings expand their view of cyclic nucleotide-gated channel function to include the regulation of synaptic transmission and mediation of the presynaptic effects of NO.

Snake venom

The antinociceptive effect of *Crotalus durissus terrificus* snake venom was reported by Picolo et al. (2000). The venom orally administered blocked carrageenin-induced hyperalgesia in the rat; the antinociception was abolished by naloxone (a δ-opioid receptor antagonist), L-NAME, and methylene blue, suggesting that both δ-opioid receptors and NO participate in the mediation of the peripheral nociceptive effect of this snake venom. Picolo and Cury (2004) noted that peripheral κ- and δ-opioid receptors were involved in the antinociceptive effect of *C. durissus terrificus* on prostaglandin E_2-induced hyperalgesia and suggested that peripheral NO, generated by nNOS, and the cyclic GMP/protein kinase pathway are responsible for molecular mechanisms of the venom effect. On the other hand, Chacur et al. (2004) noted that intraplantar injections of *Bothrops asper* snake venom components into rat hindpaws induced mechanical hyperalgesia, and the nNOS inhibitor 7-NI, but not the iNOS inhibitor L-NIL, inhibited the venom-induced hyperalgesia.

Myotoxic effects of crotoxin, the major component of the venom of *C. durissus terrificus*, were observed in rat tibialis anterior and soleus muscles; the crotoxin-induced myonecrosis was minimized by L-NAME, and partially prevented by both 7-NI and aminoguanidine, an iNOS inhibitor, indicating that NO may be an important intracellular signaling molecule that mediates crotoxin myotoxic activity (Miyabara et al. 2004). β-Bungarotoxin had a potent neurotoxic effect on cultured cerebellar granular neurons; Ca^{2+}-channel inhibitors attenuated the toxin-mediated rise in intracellular Ca^{2+}, neurotoxicity, and generation of reactive nitrogen species, and L-NAME also exhibited neuroprotection, showing that β-bungarotoxin-induced neuronal cell death may be mediated by excessive generation of NO triggered by intracellular Ca^{2+} overloading (Tseng and Lin-Shiau 2003). *B. insularis* snake venom induced mouse hindpaw edema, which was attenuated by pretreating the mice with histaminergic receptor antagonist, L-NAME, indomethacin, and dexamethasone, showing that this response may be mediated by histamine, NO, and arachidonic acid metabolites formed by cyclooxygenases (Barbosa et al. 2003).

In endothelium-intact rat aortic rings, cobra cardiotoxin induced a transient relaxation followed by a sustained contraction; removal of the endothelium or treatment with L-NAME abolished the relaxation (Ho et al. 1998). Endothelium-dependent relaxation induced by ACh was decreased by cobra cardiotoxin. Intravenous *Vipera lebetina* snake venom produced rapid cardiovascular collapse in anesthetized rats, and L-NAME attenuated the vascular hyporeactivity to phenylephrine induced by the venom in isolated rat mesenteric vascular beds, suggesting that NO or NO-like compounds may be present in the venom and involved in its hypotensive effect (Fatehi-Hassanabad and Fatehi 2004). Berythractivase and jararhagin, PIII snake venom metalloproteases with 69% homology that have different hemostatic properties, enhanced NO generation, prostacyclin production, and interleukin-8 release in human umbilical vein endothelial cells; jararhagin decreased cell viability and induced cellular apoptosis, while berythractivase did not modulate cell survival (Schattner et al. 2005). Although structurally related, these proteins induce a dissimilar generation and release of endothelial molecules that may account for their different hemorrhagic activity.

NO formed by constitutive NOS in birds

Cardiovascular system

Ali et al. (1996a) demonstrated that NADPH-diaphorase-positive and NOS-immunoreactive cell bodies and nerve fibers occurred in the wall of the bursa of Fabricius and between the lymphoid folliculus in the chicken and in the immediate vicinity of blood vessels in the bursa; the vascular endothelium was densely stained, suggesting that NO may play a role in the regulation of the secretory activity of interfollicular epithelium and the blood flow through the bursa. In studies on the expression patterns of shear stress-responsive genes Kruppel-like factor (KLF-2), endothelin-1, and eNOS in the developing cardiovascular system of chicken embryos, the expression patterns were mostly not shear stress-related in the early stages, whereas during development, the shear stress-related overlapping expression patterns of KLF-2 and eNOS became evident in the cardiac inflow and outflow tracts, aortic sac, and pharyngeal arch arteries (Groenendijk et al. 2004). The level of expression of endothelin-1 in the embryonic heart was decreased, whereas KLF-2 and eNOS were both upregulated after venous clip, suggestive of increased cardiac shear stress (Groenendijk et al. 2005).

Using a superfusion cascade system (the chicken abdominal aorta as a donor tissue and the endothelium-denuded aortic ring as a bioassay tissue), Hasegawa and Nishimura (1991) noted that the endothelium-dependent, ACh-induced relaxation was potentiated by superoxide dismutase and abolished by hemoglobin, suggesting that ACh-induced relaxation of the fowl aorta is, at least partially, mediated by a humoral factor(s) that resembles EDRF demonstrated in mammals. ACh caused relaxation mediated by NO from the endothelium also in femoral and carotid arteries isolated from the chicken embryo (days 15, 17, and 19 of incubation; Ie Noble et al. 2000).

Hypoxia-induced relaxation of the femoral artery isolated from chicken embryos was not altered by removal of the endothelium or by inhibition of NOS or cyclooxygenase inhibitors; however, relaxant responses to ACh were abolished during acute hypoxia, suggesting the inhibition by hypoxia of stimulated EDRF release (Ruijtenbeek et al. 2002). Chronic hypoxia, to which eggs of White Leghorn chickens were exposed from day 6 until day 19 of the 21-day incubation, reduced the sensitivity to ACh of isolated femoral arteries, and in the presence of L-NAME, this difference in sensitivity was no longer apparent, indicating that hypoxia reduced the sensitivity to ACh by lowering NO release (Ruijtenbeek et al. 2003). From these observations, the authors proposed that intrauterine exposure to chronic hypoxia induces changes in arterial endothelial properties that may play a role in the development of cardiovascular disease in adult life. In isolated intrapulmonary arteries from non-internally pipped embryos at 19 days of incubation and from internally and externally pipped embryos at 21 days of incubation, no developmental changes were observed in endothelium-dependent, ACh-induced, or endothelium-independent, SNP-induced relaxation; reduction of oxygen concentration from 95% to 5% suppressed relaxations induced by these agonists (Villamor et al. 2002). They concluded that chicken embryo pulmonary arteries show an NO-mediated, endothelium-dependent relaxation that is unaffected by transition to ex ovo life. In pulmonary artery rings, ACh- and A23187-induced relaxations, both of which depended on the endothelium, were smaller in broilers that are prone to pulmonary hypertension than Leghorns that are not prone to pulmonary hypertension (Martinez-Lemus et al. 1999). L-Arginine enhanced ACh-induced relaxations more in broilers than Leghorns, and relaxations to SNP did not differ between strains, suggesting that broiler pulmonary artery rings appear to have increased intrinsic tone and reduced endothelium-derived NO activity, both of which may contribute to the susceptibility of broiler chickens to pulmonary hypertension. Odom et al. (2004) noted that: (1.) In vitro hypoxia decreased NO-dependent, ACh-induced relaxation in broiler pulmonary arteries but only reduced the response to the lowest concentration of ACh used in those from Leghorns, (2.) L-Arginine supplementation failed to augment the response to ACh under hypoxia, and (3.) Relaxations to SNP were not affected by hypoxia in Leghorns but were increased by hypoxia in broilers. It appears that the increased incidence of pulmonary hypertension syndrome in broiler chickens reared under hypoxia may be associated with a hypoxia-induced reduction in the synthesis or activity of endothelial NO in the pulmonary circulation. In the isolated heart from chick embryos, L-NAME prolonged the electromechanical delay in ventricles during anoxia and delayed the recovery of this parameter during reoxygenation, while an NO donor had the opposite effect, providing evidence that an NO-dependent pathway is involved in regulation of the ventricular excitation–contraction coupling in the anoxic-reoxygenated developing heart (Maury et al. 2004).

Wang et al. (2002) provided evidence that L-NAME increases broiler pulmonary arterial pressure by inhibiting the endogenous synthesis of NO, leading to pulmonary hypertension, right ventricular hypertrophy, and the increased morbidity of pulmonary hypertension syndrome. Intravenous L-NAME also increased pulmonary arterial pressure and myocardial contractility in broiler chickens (Weidong et al. 2002). The pulmonary vascular endothelium appears to release NO that in turn reduces the pulmonary vascular resistance or attenuates myocardial contractility. Broilers with pulmonary hypertension syndrome elicited by chronic hypoxia had fewer NADPH-diaphorase-positive endothelial cells in arterioles than did the normotensive broilers, indicating that hypoxia-induced pulmonary hypertension may be associated with a decrease in endothelium-derived NO expression in pulmonary vessels

(Moreno de Sandino and Hernandez 2003). Broilers developing pulmonary hypertension syndrome triggered by exposure to low temperature exhibited diminished NOS expression in the endothelium of their pulmonary arterioles (Tan et al. 2005a). Supplemental L-arginine prevented the reduced expression of eNOS, which might contribute to the increased production of NO by the pulmonary vasculature. These authors (Tan et al. 2005b) also provided evidence that supplemental L-arginine inhibited pulmonary vascular remodeling that occurred secondary to increased pulmonary pressure. In broilers pretreated with L-NAME followed by an intravenous injection of cellulose microparticles that were entrapped within the pulmonary vasculature, increases in pulmonary arterial pressure and pulmonary vascular resistance were greater than in broilers with injected microparticles; the postinjection mortality more than doubled when L-NAME was combined with the microparticle injection, suggesting that the inhibition of NOS exposed a more dramatic increase in pulmonary arterial pressure, leading to enhanced mortality in response to microparticle injections (Wideman et al. 2005).

Maturation-dependent rises in blood pressure in chicks appeared to trigger neointimal plaque formation, which was prevented by inhibition of blood pressure increase by propranolol or via a possible increase in NO availability as seen in the L-arginine supplement group (Ruiz-Feria et al. 2004). In chickens in which arterial blood pressure was low (82.7 mmHg) and plasma norepinephrine levels (15.5 ng/ml) and angiotensin-converting enzyme (ACE) activities (87.3 U/l) were high, long-term L-NA administered with tap water increased blood pressure and decreased plasma norepinephrine and epinephrine levels and ACE activities. By contrast, in chickens, in which blood pressure was high (141.4 mmHg) and plasma norepinephrine level (1.1 ng/ml) and ACE activity (57.2 U/l) were low, long-term L-NA administration did not change blood pressure, but increased norepinephrine and ACE values, the effect of L-NA being reversed by L-arginine (Aksulu et al. 2000). These authors suggested that an L-arginine/NO pathway likely modulates the epinephrine/norepinephrine release as well as the rennin–angiotensin system in chickens.

From findings that in tanycytes of the tectal ventricle of the pigeon, NADPH-diaphorase-positive neurons surrounded and contacted the radial blood vessels that originated in the stratum griseum profundum, Meyer et al. (1994) discussed the possibility that NOS plays a role in the regulation of local blood flow in the deep tectal layers. In anesthetized pigeons, increase in choroidal blood flow elicited by electrical stimulation of the nucleus of Edinger–Westphal was attenuated by L-NAME and 7-NI, and L-arginine reversed the effects; resting choroidal blood flow was reduced and systemic blood pressure was increased after L-NAME administration (Zagvazdin et al. 1996). The authors concluded that neurally derived NO is responsible for a major component of the choroidal blood flow increase caused by Edinger–Westphal stimulation in pigeons. In retinal arteries and arterioles in anesthetized dogs, NO liberated from vasodilator nerves acts as a neurotransmitter, and the arteriolar muscle tone is regulated by the nitrergic nerve (Toda et al. 1994). Cuthbertson et al. (1997) found that the majority of the pterygopalatine ganglia neurons co-contained nNOS and VIP, and axons labeled for nNOS, NADPH-diaphorase, or VIP were traced from the pterygopalatine ganglion network to perivascular fiber plexi on orbital blood vessels in the pigeon, suggesting that pterygopalatine ganglion neurons of birds use NO and VIP to exert vasodilatory control over blood flow to and within the avian choroids.

On the basis of functional studies on cerebral and retinal arteries from dogs with or without damage of pterygopalatine ganglia, Toda et al. (1993) hypothesized that vasodilator nitrergic nerve fibers to cerebral and retinal arteries arise from the pterygopalatine ganglion. Further analyses for the origin of nitrergic nerve fibers innervating monkey and canine cerebral and ophthalmic arteries revealed that preganglionic fibers in the pterygopalatine ganglion arise from the superior salivatory nucleus in the brain stem, and the ganglion cells send off vasodilator nitrergic nerve fibers to cerebral and ocular vasculatures (Toda et al. 2000a, b; Ayajiki et al. 2000). Because himbacine, a relatively selective antagonist of M_2-type muscarinic receptors, increased the Edinger–Westphal-evoked parasympathetic vasodilatation mediated by NO in pigeon choroid, Zagvazdin et al. (2000) postulated that M_2 receptors may play a prejunctional role in downregulating the parasympathetic vasodilatation in the avian choroid. Involvement of prejunctional M_2 muscarinic receptors in the impairment of vasodilator nitrergic nerve function has been evidenced in monkey cerebral arteries (Toda et al. 1997).

Smooth muscle

In the gut of newly hatched chickens, all NADPH-diaphorase-positive neurons were also NOS-immunoreactive and all NOS-immunoreactive neurons were NADPH-diaphorase positive; the majority of VIP-immunoreactive neurons also expressed NADPH-diaphorase activity, suggesting that NO may play a role in the neural control of the gut musculature (Balaskas et al. 1995). Hiramatsu et al. (1999) also noted evidence indicating that NADPH-diaphorase can be used as a marker for NOS-containing neurons in the chicken caecum. VIP, but not NOS, immunoreactivity was detected in mucosal fibers in this species. NADPH-diaphorase neurons were also found throughout the pigeon gastrointestinal tract and they were evident in the myenteric, circular muscle, and submucous plexuses (Mirabella et al. 2000b). These authors supposed that the nitrergic nerve population of the pigeon gut is involved in muscle motility regulation as an inhibitory descending nerve pathway.

Electrical field stimulation induced relaxation of chick isolated upper esophagus treated with atropine, which was inhibited by L-NAME and tetrodotoxin; L-arginine reversed the inhibition by L-NAME, suggestive of the release of NO as an inhibitory neurotransmitter from NOS-containing neurons (Gwee et al. 1995). Similar results were also observed in the pigeon esophageal smooth muscle (Postorino et al. 1999). Vetri et al. (2004) suggested that the nonadrenergic noncholinergic inhibitory response elicited by electrical field stimulation presents apamin-sensitive, probably purine-mediated and apamin-insensitive, NO-mediated, components in pigeon esophageal smooth muscle.

In the circular muscle of isolated chicken ileum, Martin et al. (1998) obtained findings supporting the hypothesis that the inhibitory effect of cholecystokinin seems to be mediated through purinergic neurons, which in turn stimulates the release of NO and a peptide, possibly VIP. Chicken motilin caused contractions of longitudinal muscle strips of the chicken proventriculus; L-NAME potentiated contractions induced by motilin, electrical field stimulation, and a ganglionic stimulant, the potentiation by L-NAME being prevented by L-arginine (Kitazawa et al. 2002). In contrast, in the small intestine, L-NAME and L-arginine were both ineffective in modifying the motilin-induced contraction. Therefore, these authors suggested that NOS inhibition seems to potentiate the contractile response to motilin, electrical field stimulation, and the ganglionic stimulant in the chicken proventriculus through reduction of endogenous NO-mediated prejunctional inhibition on neurogenic ACh release; however, NOS inhibition does not modify the myogenic action of motilin in gastric and intestinal smooth muscles.

In chickens that were implanted with electrodes for electromyography in the stomach, cholecystokinin-induced inhibition of gastric motility was blocked by vagotomy and L-NAME, suggesting that the gastric response to this peptide is vagally mediated, mainly by the NO system (Martinez et al. 1993). In chickens, like rats, with implanted electrodes for electromyography in the small intestine, Rodriguez-Membrilla et al. (1995) provided evidence that an inhibition of NO synthesis seems to be involved in the induction of the fasting motor pattern, whereas an increase of NO mediates the occurrence of the fed pattern.

In the duck ureter, nerve lesioning experiments showed that the majority of nitrergic innervation was extrinsic in origin; nitrergic adventitial neurons projected caudocranially, whereas NOS-immunoreactive and NOS-/VIP-immunoreactive intramural neurons projected craniocaudally (Mirabella et al. 2000a). These authors suggested that in birds, the nitrergic innervation plays a role in ureteral functions, such as epithelial mucosecretion, muscular motility, and the closing and/or opening of the ureteral papilla. According to Sann (1998), the relative numbers of neuropeptide Y-, NOS-, calbindin-, and VIP-positive neurons were 57%, 28%, 14%, and 7%, respectively, in autonomic ganglia associated with the chicken ureter. All NADPH-diaphorase-positive neurons expressed NOS immunoreactivity.

In the upper segments of the hen oviduct, NADPH-diaphorase- and VIP-positive nerve structures were numerous in the intermuscular and mucosal layers, and they were represented by fibers and cell bodies showing a perivascular distribution; in the distal zone of the oviduct and in the vagina, nitrergic and VIP-positive nerve fibers were widely diffused in the circular muscle (Costagliola et al. 1997). NADPH-diaphorase-positive neurons were mainly distributed around the arterioles of the intermuscular tissue in the upper oviduct of the pigeon; in addition, these positive neurons were also seen in the smooth muscle layers and lamina propria in the lower oviduct (uterus and vagina). By double staining, most of the neurons showed colocalization of NADPH-diaphorase and acetylcholinesterase in the uterus and vagina (Atoji et al. 2000). Nitrergic neurons were demonstrated to make up a large subpopulation of intrinsic neurons that were closely associated with a cholinergic system, thus suggesting that NO and ACh could be used to modify the relaxation of the avian oviduct. Colocalization of NOS, VIP, and pituitary adenylate cyclase-activating polypeptide (PACAP) were detected in an intrinsic neuronal subpopulation in the vaginal segment of the chicken oviduct; electrical field stimulation of isolated vaginal muscle strips induced relaxations that were blocked by tetrodotoxin and reduced by L-NA, supporting a hypothesis that NO, VIP, and PACAP exert an inhibitory action on muscle strips of the vaginal part of the chicken oviduct (Costagliola et al. 2004).

Central nervous system

Bruning (1993) detected localization of NADPH-diaphorase in the brain of the chicken and found intensely stained neurons and fibers in most parts of the telencephalon. His conclusion was that the pattern of expression of NADPH-diaphorase and possibly NOS within avian and mammalian brain might be largely conserved. Later, Bruning et al. (1994b) found that various structures in the chick brain, where NADPH-diaphorase staining has been suggested to contain NOS, were not always nNOS-immunoreactive; nNOS was sparsely present in the hyperstriatum ventrale, providing evidence against the involvement of NO in certain forms of learning and memory processes known to occur in this region. There was evidence not supporting the hypothesis that learning-related changes in the intermediate and medial part of the hyperstriatum ventrale depend on NO acting as a retrograde neuronal messenger (Ambalavanar et al. 1994).

NADPH-diaphorase-containing neurons were present in olfactory, visual, auditory, and somatosensory tracts of the pigeon brain, suggesting that NO modulates sensory transmission in avian CNS (Atoji et al. 2001). Using pigeon brain slices and microiontophoresis, Hu et al. (2001) showed that 60% of isthmo-optic cells received glutamatergic afferents, 20%

received nitrergic afferents, and 20% of others received both glutamatergic and nitrergic afferents from the tectum; in the last group, both glutamate and NO seemed to co-release from the same tecto-isthmo-optic terminals. In the developing chicken neural tube, where NO apparently regulated cell-cycle progression, Traister et al. (2002) provided evidence that high NO levels promote entry into the S phase basally, whereas low levels of NO facilitate entry into mitosis apically.

In the pigeon, where the ciliary ganglion is known to control the lens and pupil function and the choroidal vasculature, about one-third of ganglion cells were shown to be densely NADPH-diaphorase-positive, while in the cat and monkey, only a small number (less than 2%) of the ganglion cells were densely stained (Sun et al. 1994). The presence of NADPH-diaphorase-positive cells in the avian and mammalian ciliary ganglia suggests that NO may be used for intercellular communication in this ganglion, or in light of the known importance of NO in vascular control, some of these positive neurons may participate in the control of choroidal vasodilatation. Cuthbertson et al. (1999) provided evidence indicating that nNOS is present in the Edinger–Westphal nucleus-arising preganglionic endings on choroidal and ciliary neurons in the pigeon ciliary ganglion. Torrao and Britto (2004) found evidence that NOS expression in visual structures of the adult chick brain is downregulated by the retinal innervation; alternatively, the increase of NOS expression observed after retinal removal could be an indicator of a role of the NO system in plasticity processes. In the Japanese quail, nNOS-immunoreactive pterygopalatine neurons were surrounded by somatostatin-positive nerve fibers, and in the choroid, nNOS-positive intrinsic choroidal neurons were closely apposed by somatostatin-immunoreactive nerve fibers, suggesting that somatostatin may regulate the production of NO in pterygopalatine neurons and choroidal neurons and is involved in neuronal circuits regulating ocular homeostasis (Schrodl et al. 2005). In the chick ciliary ganglion of post-hatched birds, SNP and 8-bromo-cyclic GMP increased the efficacy of synaptic transmission, and L-NAME added prior to the tetanic stimulation of the preganglionic nerves reduced the long-term potentiation but did not change the post-tetanic potentiation (Scott and Bennett 1993). A component of long-term potentiation was shown to be due to the release of NO in the ganglion, as blocking the synthesis of NO with L-NAME decreased the potentiation, suggesting that endogenous NO is involved in either the initiation or maintenance phase of long-term potentiation (Lin and Bennett 1994).

Although, in contrast to mammals, preganglionic neurons labeled only weakly for NADPH diaphorase in the spinal cord of the pigeon, choline acetyltransferase and NADPH diaphorase systems of birds seemed to be largely similar to those of the mammalian spinal cord (Necker 2004).

NOS in chicken cerebellum was dependent upon Ca^{2+} and calmodulin and required co-factors such as NADPH, FAD, and FMN for its full expression, suggesting that chicken cerebellum NOS has similar properties to those of mammalian nNOS (Yamakawa et al. 1997).

Memory

Holscher and Rose (1992) showed that injection of L-NA prior to training resulted in amnesia for the passive avoidance task in the chick; the amnesia could be overcome by injecting L-arginine along with the inhibitor. Rickard et al. (1998) supported NO-dependent memory formation in the day-old chick. Injection of L-NA into one intermediate medial hyperstriatum ventrale alone (left or right), an area that is of crucial importance in learning in the chick, proved to be sufficient to produce amnesia; diffusion of the NOS inhibitor into the untreated hemisphere was ruled out by injecting it with L-arginine (Holscher and Rose 1993). 7-NI, an nNOS inhibitor, had an amnesic effect in the avoidance task; thus, the possible involvement of vascular action in amnesia may be excluded (Holscher 1994). The author hypothesized that NO is a neuronal transmitter that is important in processes of synaptic plasticity and learning in birds.

In contrast, Rickard et al. (1999) noted that administration of a specific nNOS or iNOS inhibitor had no effect on memory retention in the chick trained on a single-trial passive-avoidance task, whereas the eNOS inhibitor or nonspecific NOS inhibitors impaired long-term memory formation, indicating that eNOS may play a role in memory formation for this task in birds. Later, these authors suggested that both eNOS and nNOS isoforms may be essential for long-term memory consolidation of the avoidance task in the chick (Rickard and Gibbs 2003a). Von Bartheld and Schober (1997) postulated from morphological and neurochemical studies on the chick brain that NOS-containing presumptive local circuit neurons in the lobus parolfactorius in the telencephalon are the most likely source of NO involved in learning of passive avoidance tasks. Administration of L-NAME into the lobus parolfactorius impaired memory formation for an avoidance task in the chick; the effective administration time was hemisphere-dependent, requiring left hemisphere administration around the time of training and right hemisphere administration between 15 and 25 min post-training, indicating that localized NO activity in each hemisphere of the chick brain may be necessary for the consolidation of memory for this task (Rickard and Gibbs 2003b).

The field response evoked by local electrical stimulation in the intermediate and medial part of the hyperstriatum ventrale in an in vitro slice preparation from the chick brain was depressed by NO donors but not affected by ODQ, a soluble guanylyl cyclase inhibitor; long-term depression by NO donors was eliminated in the presence of either a selective adenosine A_1-receptor antagonist or an ADP–ribosyltransferase inhibitor, suggesting that NO may be acting presynaptically in a synergistic fashion with the adenosine A_1 receptor to depress transmitter release (Barcellos et al. 2000). Mono ADP-ribosylation inhibition seems to be required for the maintenance of long-term memory in day-old chicks (Edwards and Rickard 2002). Pharmacological inhibition of guanylyl cyclase resulted in two transitory memory retention deficits centered around 40 and 120 min post-training, respectively, while inhibition of protein kinase G elicited a single temporary retention loss centered at 120 min post-training; these agents likely inhibit memory retrieval rather than formation (Edwards et al. 2002). It appears that guanylyl cyclase mediates two memory retrieval processes, the latter of which would be dependent on protein kinase G and that since inhibition of NO results in a permanent retention loss, NO is required for memory formation through guanylyl cyclase-independent processes.

Edwards and Rickard (2005) obtained results suggesting that physiological levels of peroxynitrite may be required for the consolidation of long-term memory in the chick and that NO acts to facilitate memory formation through the production of peroxynitrite. In chicks trained on a passive avoidance task, inhibition of CO production using zinc deuteroporphyrin resulted in retention loss, as observed when the soluble guanylyl cyclase activity was inhibited, and the memory deficit was abolished by heme oxygenase agonist hemin, supporting a role for endogenous CO in memory processing, possibly through activation of guanylyl cyclase (Cutajar et al. 2005).

Behavior

Intraperitoneal or intracerebroventricular injections of L-NAME inhibited the food intake of broiler chickens; the effect of L-NAME was attenuated by intracerebroventricular administration of L-arginine, suggesting that central NO may control feeding behavior in the chicken (Choi et al. 1994). L-NAME also inhibited food intake stimulated by clonidine, an α_2-adrenoceptor agonist, indicating the possibility that NO interacts with adrenergic neurons in the CNS to modulate feeding behavior in the chicken (Choi et al. 1995). Both 7-NI and L-NAME produced phencyclidine-like behavioral effects and catalepsy in pigeons, the former being more effective; pretreatment with L-arginine prevented these effects (Jewett et al. 1996).

Eyes

Morgan et al. (1994) noted that NADPH-diaphorase-positive elements in the chick retina degenerated after destruction of the isthmo-optic nucleus or tract, indicating that the centrifugal projection to the retina in birds appears to use NO as a messenger or transmitter. In the retina of hatched chicks, NADPH-diaphorase and nNOS-immunolabeling were present in amacrine cells, ganglion cells, efferent fibers, efferent target cells, and neural processes in plexiform layers, whereas diaphorase alone was detected in photoreceptor ellipsoids and Muller cells (Fischer and Stell 1999). In addition, nNOS was detected in axon bundles and innervation to blood vessels to the choroid, whereas stromal and endothelial cells in the choroid, and retinal pigmented epithelium contained NADPH diaphorase but not nNOS. The presence of eNOS immunoreactivity was reported in Muller cells of chicken as well as other vertebrates (Haverkamp et al. 1999). There was evidence suggesting that intrinsic choroidal neurons received a sympathetic input that might modulate their nitrergic effects upon vascular and nonvascular smooth muscle fibers in the choroid (Schroedl et al. 2001). Using a neural tracer cholera toxin, which was injected in the ocular globe of the post-hatched chick, Gardino et al. (2004) found that acetylcholinesterase activity was present mainly in cholera toxin-labeled cell bodies of the isthmo-optic nucleus and the ectopic region, while NADPH-diaphorase histochemistry was present in the neutrophil and sparse cell bodies inside of the isthmo-optic nucleus and in ectopic neurons, which were not cholera toxin-positive, supporting the idea that these two neurochemical systems are present in distinct neuronal populations in the centrifugal visual system. Rios et al. (2000) demonstrated development of nitrergic neurons in the chick embryo retina by using the NADPH-diaphorase reaction. Distribution of intrinsic choroidal neurons in 12 avian species was compared by NADPH-diaphorase histochemistry or nNOS-immunocytochemistry (Schroedl et al. 2004).

Fujii et al. (1998) noted that expression of iNOS mRNA was highest in the chick retina-retinal pigment epithelium-choroid, followed by the expression of nNOS mRNA, and the eNOS mRNA expression was faint; however, the iNOS protein level was only slightly higher than the levels of the nNOS and eNOS. nNOS alone was found in the outer nuclear layer. It was suggested that NOS isoforms may be differentially involved in the mechanisms regulating the posterior eye tissues, including myopic eye growth. BH_4 is an essential cofactor for the biosynthesis of NO. Holdengreber et al. (2002) demonstrated that 6-pyruvoyltetrahydropterin synthase was already expressed in the undifferentiated and proliferating retina at E7; and at E11 the enzyme was expressed in photoreceptors, amacrine cells, displaced amacrine cells, and ganglion cells, as well as in the plexiform layers, in which synaptic connections take place.

Incubation of dark-adapted and dark-maintained chicken retinas with SNP led to a NO-specific suppression of the arylalkylamine *N*-acetyltransferase activity, which responds to light and reflects the changes in retinal melatonin synthesis; cyclic GMP analogs and cyclic GMP-specific phosphodiesterase inhibitor zaprinast also suppressed this enzyme activity, while an inhibitor of soluble guanylyl cyclase blocked the effect of SNP (Wellard and Morgan 2004). During and after spreading depression wave propagation, administration of SNP or nitroglycerin reduced the magnitude of intrinsic optical signals in the chicken retina, observed via a microscope (Dahlem and Hanke 2005).

Myopia

In 6-day-old chicks, the eyes of which were occluded for 6 days with goggles, the refractive errors and axial length were less affected in left eyes injected with L-NAME compared to control (right) eyes; changes in electroretinogram were reversible, except in eyes injected with the highest concentrations of L-NAME used (Fujikado et al. 1997). The chicks, injected with L-NAME and reared without occlusion, had normal refractive values. The injection of L-NAME before occlusion of developing chick eyes appears to lead to reversible modifications in retinal function and inhibits the development of form-deprivation myopia. The injection of L-NAME into developing chick eyes that were then covered with a –16-D lens resulted in a modification of retinal function and an inhibition of the development of myopia, suggesting that NO modulates a common retinal pathway that leads to both lens-induced myopia and form-deprivation myopia (Fujikado et al. 2001). When the retina of chick eyes is exposed to myopic defocus, the choroid thickens, pushing the retina forward; conversely, when the eye is exposed to hyperopic defocus, the choroid thins. Nickla and Wildsoet (2004) found that L-NAME injected intravitreally inhibited choroidal thickening in both previously form-deprived eyes and eyes wearing +15-D lenses, suggesting that NO may play a role in modulating choroidal thickness. In myopic eyes in form sense-deprived chicks, zinc drops inhibited the elongation of axis oculi and increased the diopters in myopia; the activities of NOS and superoxide dismutase and the content of NO in retinopigmental epithelium-choroid homogenate of the eyes in groups supplemented with zinc were increased compared with the model group, indicating that zinc may be used to prevent and treat myopia (Xu et al. 2001a). Nacreous powder, containing calcium and many mineral elements as well as being rich in amino acids, had the ability to prevent the elongation of the axis oculi and change the diopters of form sense-deprived model chicks, possibly due to increased activities of NOS and superoxide dismutase (Xu et al. 2001b).

Others

In the medulla of the chick thymus, NADPH-diaphorase-positive cells were present as clusters, which were closely associated with the thymic cysts; some cells were seen in close proximity to blood vessels, suggesting a modulatory influence of NO over the activities of these cells (Gulati et al. 1993). Intracellular localization of NADPH-diaphorase was detected by electron microscopy in the chick thymic medulla (Gulati et al. 1995). Further studies by these authors (Gulati et al. 1997) demonstrated that out of VIP, CGRP, substance P, and neuropeptide Y, VIP-immunoreactive nerves in the chick thymus appeared to be the most abundant, and VIP and NADPH-diaphorase were colocalized in the neuronal structures. Hiramatsu and Ohshima (1994) noted that NADPH-diaphorase colocalized with VIP and galanin in ganglion cells of the chicken pancreas. On the basis of distribution patterns of NADPH-diaphorase and NOS, Ali et al. (1996b) suggested that NO may play a role in the regulation of the secretory activity of and the blood flow through ultimobranchial glands of the chicken.

In healthy non-mammal skeletal muscle, that is, of birds, reptiles and fishes, caveolin-3 immunoreactivity was lacking in the sarcolemma as was α-sarcoglycan; the other nNOS-dystrophin complex components were either present or absent, suggesting that functions of caveolin and sarcoglycan are subserved by other components of the nNOS-dystrophin complex (Gossrau 1998). Puttmann et al. (2005) provided evidence from studies using 5- to 6-day chick embryos and mouse skeletal C2C12 myoblasts that in target skeletal muscle cells the NOS/NO system is controlled by motoneuron contacts, and endogenous NO signaling in myotubes may be essential during synapse formation and plasticity of the neuromuscular system.

There was evidence supporting the hypothesis that NO plays a key role in regulating follicular development, ovulatory mechanisms, and egg production in Japanese quail (Manwar et al. 2006).

Ontogenesis of constitutive NOS in mammals and non-mammal vertebrates

Mammals

Derer and Derer (1993) noted that NADPH-diaphorase-reactive neurons first appeared between embryonic days 15 and 16 (E15 and E16) and were confined to the front-lateral aspect of the incipient corpus striatum and cortical subplate, but were not present in the cortical plate in the mouse forebrain; until birth these neurons differentiated progressively. In the mouse claustrum, densely nNOS-stained neurons were mostly GABA-immunoreactive and appeared to originate in the subpallium; they were first seen at E17.5 (Guirado et al. 2003). The basomedial nucleus exhibited the highest nNOS immunoreactivity in the mouse amygdalar basolateral complex, already distinguished from E15.5; each nucleus of the amygdalar basolateral complex displayed a distinct nNOS expression pattern, which was established during the ontogenesis with minor changes in the adult (Olmos et al. 2005).

In the rat olfactory bulb, a scant number of NADPH-diaphorase-reactive neurons were first present at E21 in an immature phenotype, and from E21 to postnatal day 7 (P7), neural processes seemed to contact blood vessels (Samama and Boehm 1996). Endothelial cells showed a diffuse and faint staining at all stages from E15 to P30; moreover, patches of high NADPH-diaphorase staining were present on blood vessels. By the 19th day of embryonic development in the rat heart, eNOS was present not only in the endocardium but also in the intima of numerous small vessels; in the neonatal rat heart, vascular eNOS predominated, with extensive reaction product in arteries, veins, and numerous capillary size vessels within the myocardium (Ursell and Mayes 1996). Terada et al. (2001) found that a few neurons positive for nNOS were first detected at E15 in the hypothalamus and pons of the rat brain, and many positive cells became detectable in the thalamus at E17. In the forebrain and midbrain, the distribution pattern of nNOS-containing neurons was fundamentally completed by E19. In the rat diaphragm, NOS activity, nNOS and eNOS protein, and mRNA expressions were markedly increased during late gestational and early postnatal periods; however, expression of both isoforms declined progressively thereafter (El Dwairi et al. 1998). Immunostaining revealed extensive nNOS expression at the sarcolemma, whereas eNOS expression was limited to the endothelium.

NOS in the homogenate of pig enterocytes required the presence of NADPH and BH_4 for activation, but did not appear to necessitate exogenous FAD and FMN; the enzyme activity was not affected by added Ca^{2+} or ethyleneglycoltetraacetic acid (EGTA) and was inhibited by L-NMMA and L-NA (M'Rabet-Touil et al. 1993). The NOS activity was not detectable in enterocytes isolated at birth and increased slightly in suckling animals.

To date in the literature, nNOS appears at E15–E21 in the brain and eyes of rats, whereas eNOS is first observed in blood vessels at E15–E19. Although ontogenesis of these isoforms was histologically determined to be almost identical to that in mammals, developmental establishment of the functional roles of NO formed by nNOS and eNOS has not been elucidated. According to Donald and Broughton (2005), anatomical evidence has already been found for nNOS in fish, amphibian, and lizard neurons, while eNOS first appears in reptiles but not in arteries and veins of various species of fishes and amphibians. However, there is some evidence suggesting the functional role of NO derived from the endothelium in the fish (Fritsche et al. 2000) and frog (Knight and Burnstock 1996). In a variety of fishes, endothelium-dependent vasodilators are functionally determined as being mediated by prostaglandins (see Table 1).

Non-mammal vertebrates

Villani (1999) discovered that the first appearance of NADPH-diaphorase-positive neurons occurred 20 h after fertilization in the olfactory placodes and in the neuronal tube in the diencephalons and hindbrain of the cichlid fish *T. mariae*. The earliest NOS-positive neurons were observed in the inferior reticular nucleus in the caudal rhombencephalon at embryonic stage 30 in the urodele amphibian (Moreno et al. 2002a, b). In the gastrointestinal tract of the axolotl, NOS immunoreactivity did not appear before E6 when it was found infrequently in nerve fibers (Maake et al. 2001). The first NOS-immunoreactive nerve fibers were detectable at the first juvenile state (day 64 after hatching) mainly in submucosal blood vessels in the axolotl gastrointestinal tract (Badawy and Reinecke 2003). The early appearance of VIP-immunoreactive (day 3) and PACAP-immunoreactive nerve fibers (day 6) in ontogeny may suggest important roles of these peptides in the control of gastrointestinal activity, while VIP/PACAP/NOS-positive fibers that appear late may be involved in the regulation of submucosal blood flow.

Nichol et al. (1995) found that NADPH-diaphorase activity first appeared in avian ciliary neurons at E10; the number of positive neurons appeared maximal at this age and thereafter declined. NADPH-diaphorase was first expressed in the epithelial cells located at the corticomedullary junction of the chicken thymic rudiment on day 13 of incubation; the number of labeled cells increased from E13 to E21, and these fibers formed the perivascular plexus by E21 (Gulati et al. 1998). In the White Leghorn chick, NADPH-diaphorase staining first appeared in the isthmic region of the midbrain at E10; the staining intensity of the isthmic nuclei dramatically increased between day 12.5 and day 13 of incubation and reached its maximum at E17 and then slowly decreased until the end of the experimental period (120 days post-hatched; Xiao et al. 2000). The transient appearance of NOS is probably necessary for neuronal differentiation or the establishment of synaptic connection in the isthmic nuclei. In the pigeon, choline acetyltransferase-positive dorsal horn neurons appeared only on the day of hatching (E18), whereas NOS-positive neurons were first detected at E14 (Necker 2005). Between incubation days 12 and 19 in myenteric ganglia in the small intestine and colon of the chick embryo, the NADPH-diaphorase-positive nerve cell density decreased with age; however, the number of nitrergic cells per ganglion was constant or even increased (Bagyanszki et al. 2000).

Conclusion

The developmental history of cNOS and physiological roles of NO in non-mammals have been summarized in this article. It was a surprise to see that yeast has an NO synthesizing enzyme that resembles the one in animal tissues, since NO has been regarded as an intercellular messenger or neurotransmitter. However, a considerable debate has recently arisen regarding an NOS inside mitochondria that affords organelle-based regulatory mechanisms for NO synthesis. Already in primitive invertebrates, NO synthesized by cNOS with properties similar to those in mammals contributes to the regulation not only of nerve functions but also tissue development, metabolism, and sensations. EDRF in fish vasculature is mainly prostaglandins, but efferent nitrergic innervation seems to be established in the gastrointestinal tract and possibly blood vessels in fishes. A phylogenic appearance of an endothelium-derived NO-sensitive mechanism for vasodilatation may be associated with the transition from water to air respiration. Birds appear to have established a cNOS-derived NO/cyclic GMP system like that seen in mammals. Efforts in the future to overcome our lack of knowledge about the functional roles of constitutively formed NO in non-mammalian tissues would provide us with new exciting insights into as yet little-explored segments of NO biology.

References

Agnisola C (2005) Role of nitric oxide in the control of coronary resistance in teleosts. Comp Biochem Physiol A Mol Integr Physiol 142:178–187

Aksulu HE, Bingol I, Karatas F, Sagmanligil H, Ustundag B (2000) Changes in plasma angiotensin-converting enzyme activity and noradrenaline responses to long-term nitric oxide inhibition vary depending on their basal values in chickens. Physiol Res 49:175–182

Ali SM, Chan AS, Leong SK (1996a) Histochemical and immunohistochemical localization of nitrergic neuronal and non-neuronal cells in the bursa of Fabricius of the chicken. Cell Tissue Res 285:273–279

Ali SM, Chan AS, Leong SK (1996b) Localization of nitrergic neuronal and non-neuronal cells in the ultimobranchial glands of the chicken. Anat Embryol (Berl) 193:161–168

Allaerts W, De Vente J, Markerink-Van Ittersum M, Tuinhof R, Roubos EW (1998) Topographical relationship between neuronal nitric oxide synthase immunoreactivity and cyclic 3′,5′-guanosine monophosphate accumulation in the brain of the adult Xenopus laevis. J Chem Neuroanat 15:41–56

Allaerts W, Koopman WJ, Verlaan BP, Buzzi M, Steerenberg PA (2000) Endogenous production of nitric oxide and effects of nitric oxide and superoxide on melanotrope functioning in the pituitary pars intermedia of Xenopus laevis. Nitric Oxide 4:15–28

Ambalavanar R, McCabe BJ, Horn G (1994) NADPH diaphorase (nitric oxide synthase) in a part of the chick brain involved in imprinting. Brain Res 644:160–163

Andersson KE, Alm P (2000) Neurogenic nitric oxide and the lower urinary tract. In: Toda N, Moncada S, Furchgott RF, Higgs EA (eds) Nitric oxide and the peripheral nervous system. Portland Press, London, pp 77–97

Andersson KE, Garcia Pascual A, Forman A, Tottrup A (1991) Non-adrenergic, non-cholinergic nerve-mediated relaxation of rabbit urethra is caused by nitric oxide. Acta Physiol Scand 141:133–134

Ando H, Shi Q, Kusakabe T, Ohya T, Suzuki N, Urano A (2004) Localization of mRNAs encoding alpha and beta subunits of soluble guanylyl cyclase in the brain of rainbow trout: comparisons with the distribution of neuronal nitric oxide synthase. Brain Res 1013:13–29

Angus JA, Cocks TM (1989) Endothelium-derived relaxing factor. Pharmacol Ther 41:303–351

Araujo DA, Cordeiro MN, Diniz CR, Beirao PS (1993) Effects of a toxic fraction, PhTx2, from the spider Phoneutria nigriventer on the sodium current. Naunyn Schmiedebergs Arch Pharmacol 347:205–208

Atoji Y, Mizutani K, Yamamoto Y, Suzuki Y (2000) Innervation of the pigeon oviduct: correlation of NADPH diaphorase with acetylcholinesterase, tyrosine hydroxylase, and neuropeptides. Auton Neurosci 84:1–7

Atoji Y, Yamamoto Y, Suzuki Y (2001) Distribution of NADPH diaphorase-containing neurons in the pigeon central nervous system. J Chem Neuroanat 21:1–22

Axelsson M, Olsson C, Gibbins I, Holmgren S, Franklin CE (2001) Nitric oxide, a potent vasodilator of the aortic anastomosis in the estuarine crocodile, Crocodylus porous. Gen Comp Endocrinol 122:198–204

Ayajiki K, Tanaka T, Okamura T, Toda N (2000) Evidence for nitroxidergic innervation in monkey ophthalmic arteries in vivo and in vitro. Am J Physiol 279:H2006–H2012

Badawy G, Reinecke M (2003) Ontogeny of the VIP system in the gastro-intestinal tract of the Axolotl, Ambystoma Mexicanum: successive appearance of co-existing PACAP and NOS. Anat Embryol (Berl) 206:319–325

Bagyanszki M, Roman V, Fekete E (2000) Quantitative distribution of NADPH-diaphorase-positive myenteric neurons in different signets of the developing chicken small intestine and colon. Histochem J 32:679–684

Balaskas C, Saffrey MJ, Burnstock G (1995) Distribution and colocalization of NADPH-diaphorase activity, nitric oxide synthase immunoreactivity, and VIP immunoreactivity in the newly hatched chicken gut. Anat Rec 243:10–18

Barbosa AM, de Amaral RO, Teixeira DF, Hyslop S, Cogo JC (2003) Pharmacological characterization of mouse hind paw oedema induced by *Bothrops insularis* (jararaca ilhoa) snake venom. Toxicon 42:515–523

Barcellos CK, Bradley PM, Burns BD, Webb AC (2000) Effects of nitric oxide release in an area of the chick forebrain which is essential for early learning. Brain Res Dev Brain Res 121:79–87

Barnes RD, Ward LE, Frank KP, Tyce GM, Hunter LW, Rorie DK (2001) Nitric oxide modulates evoked catecholamine release from canine adrenal medulla. Neuroscience 104:1165–1173

Baydoun AR, Mann GE (1994) Selective targeting of nitric oxide synthase inhibitors to system y+ in activated macrophages. Biochem Biophys Res Commun 200:726–731

Bicker G (2005) STOP and GO with NO: nitric oxide as a regulator of cell motility in simple brains. Bioessays 27:495–505

Blute TA, Mayer B, Eldred WD (1997) Immunocytochemical and histochemical localization of nitric oxide synthase in the turtle retina. Vis Neurosci 14:717–729

Blute TA, Velasco P, Eldred WD (1998) Functional localization of soluble guanylate cyclase in turtle retina: modulation of cGMP by nitric oxide donors. Vis Neurosci 15:485–498

Blute TA, De Grenier J, Eldred WD (1999) Stimulation with N-methyl-D-aspartate or kainic acid increases cyclic guanosine monophosphate-like immunoreactivity in turtle retina: involvement of nitric oxide synthase. J Comp Neurol 404:75–85

Blute TA, Lee MR, Eldred WD (2000) Direct imaging of NMDA-stimulated nitric oxide production in the retina. Vis Neurosci 17:557–566

Blute TA, Strang C, Keyser KT, Eldred WD (2003) Activation of the cGMP/nitric oxide signal transduction system by nicotine in the retina. Vis Neurosci 20:165–176

Bodnarova M, Martasek P, Moroz LL (2005) Calcium/calmodulin-dependent nitric oxide synthase activity in the CNS of Aplysia californica: biochemical characterization and link to cGMP pathways. J Inorg Biochem 99:922–928

Bordieri L, Persichini T, Venturini G, Cioni C (2003) Expression of nitric oxide synthase in the preoptic-hypothalamo-hypophyseal system of the teleost Oreochromis niloticus. Brain Behav Evol 62:43–55

Bredt DS, Snyder SH (1990) Isolation of nitric oxide synthetase, calmodulin-requiring enzyme. Proc Natl Acad Sci USA 87:682–685

Bredt DS, Hwang PM, Glatt CE, Lowenstein C, Reed RR, Snyder SH (1991) Cloned and expressed nitric oxide synthase structurally resembles cytochrome P-450 reductase. Nature 351:714–718

Broughton BR, Donald JA (2002) Nitric oxide regulation of the central aortae of the toad Bufo marinus occurs independently of the endothelium. J Exp Biol 205:3093–3100

Broughton BR, Donald JA (2005) Nitric oxide of large veins in the toad Bufo marinus. J Comp Physiol [B] 175:157–166

Brunelli E, Perrotta I, Talarico E, Tripepi S (2005) Localization of two nitric oxide synthase isoforms, eNOS and nNOS, in the skin of Triturus italicus (Amphibia, Urodela) during development. Comp Biochem Physiol A Mol Integr Physiol 142:249–255

Bruning G (1993) Localization of NADPH-diaphorase in the brain of the chicken. J Comp Neurol 334:192–208

Bruning G, Wiese S, Mayer B (1994a) Nitric oxide synthase in the brain of the turtle Pseudemys scripta elegans. J Comp Neurol 348:183–206
Bruning G, Funk U, Mayer B (1994b) Immunocytochemical localization of nitric oxide synthase in the brain of the chicken. Neuroreport 5:2425–2428
Bruning G, Hattwig K, Mayer B (1996) Nitric oxide synthase in the peripheral nervous system of the goldfish, Carassius auratus. Cell Tissue Res 284:87–98
Bult H, Boeckxstaens GE, Pelckmans PA, Jordaens FH, Van Maercke YM, Herman AG (1990) Nitric oxide as an inhibitory non-adrenergic non-cholinergic neurotransmitter. Nature 345:346–347
Burleson ML, Milsom WK (1993) Sensory receptors in the first gill arch of rainbow trout. Respir Physiol 93:97–110
Burnett AL, Lowenstein CJ, Bredt DS, Chang TSK, Snyder SH (1992) Nitric oxide: a physiologic mediator of penile erection. Science 257:401–403
Cao L, Eldred WD (2001) Subcellular localization of neuronal nitric oxide synthase in turtle retina: electron immunocytochemistry. Vis Neurosci 18:949–960
Cao L, Eldred WD (2003) Inhibitors of nitric oxide synthase block carbon monoxide-induced increases in cGMP in retina. Brain Res 988:78–83
Cartier F, Do-Rego JL, Remy-Jouet I, Fournier A, Vaudry H, Delarue C (2001) Evidence for the involvement of nitric oxide in the control of steroid secretion by the frog adrenal gland. J Steroid Biochem Mol Biol 77:251–259
Castel H, Jegou S, Tonon MC, Vaudry H (2000) Regulation of the GABAA receptor by nitric oxide in from pituitary melanotrophs. Endocrinology 141:345–360
Cayre M, Malaterre J, Scotto-Lomassese S, Holstein GR, Martinelli GP, Forni C, Nicolas S, Aouane A, Strambi C, Strambi A (2005) A role for nitric oxide in sensory-induced neurogenesis in a adult insect brain. Eur J Neurosci 21:2893–2902
Chacur M, Gutierrez JM, Milligan ED, Wieseler-Frank J, Britto LR, Maier SF, Watkins LR, Cury Y (2004) Snake venom components enhance pain upon subcutaneous injection: an initial examination of spinal cord mediators. Pain 111:65–76
Champlin DT, Truman JW (2000) Ecdysteroid coordinates optic lobe neurogenesis via a nitric oxide signaling pathway. Development 127:3543–3551
Choi YH, Furuse M, Okumura J, Denbow DM (1994) Nitric oxide controls feeding behavior in the chicken. Brain Res 654:163–166
Choi YH, Furuse M, Okumura J, Denbow DM (1995) The interaction of clonidine and nitric oxide on feeding behavior in the chicken. Brain Res 699:161–164
Church JE, Hodgson WC (2002a) Adrenergic and cholinergic activity contributes to the cardiovascular effects of lionfish (Pterois volitans) venom. Toxicon 40:787–796
Church JE, Hodgson WC (2002b) The pharmacological activity of fish venom. Toxicon 40:1083–1093
Cioni C, Francia N, Fabrizi C, Colasanti M, Venturini G (1998) Partial biochemical characterization of nitric oxide synthase in the caudal spinal cord of the teleost Orechromis niloticus. Neurosci Lett 253:68–70
Cioni C, Bordieri L, De Vito L (2002) Nitric oxide and neuromodulation in the caudal neurosecretory system of teleosts. Comp Biochem Physiol B Biochem Mol Biol 132:57–68
Cleeter MW, Cooper JM, Darley-Usmar VM, Moncada S, Schapira AH (1994) Reversible inhibition of cytochrome c oxidase, the terminal enzyme of the mitochondrial respiratory chain, by nitric oxide. Implications for neurodegenerative diseases. FEBS Lett 345:50–54
Closs EI, Basha FZ, Habermeier A, Forstermann U (1997) Interference of L-arginine analogues with L-arginine transport mediated by y+ currier hCAT-2B. Nitric Oxide 1:65–73
Cole AG, Mashkournia A, Parries SC, Goldberg JI (2002) Regulation of early embryonic behavior by nitric oxide in the pond snail Helisoma trivolvis. J Exp Biol 205:3143–3152
Costagliola A, Mayer B, Vittoria A, Carrese E, Lamanna C, Cecio A (1997) NADPH-diaphorase-, nitric oxide synthase- and VIP-containing nerve structures in the hen oviduct: a histochemical and immunohistochemical study. Arch Histol Cytol 60:245–256
Costagliola A, De Man JG, Majewski M, Lakomy M, Cecio A, Robberecht P, Pelckmans PA, Adriaensen D, Timmermans JP (2004) Coexistence of non-adrenergic non-cholinergic inhibitory and excitatory neurotransmitters in a large neuronal subpopulation in the vaginal segment of the chicken oviduct. Auton Neurosci 112:37–48
Cristino L, Pica A, Della Corte F, Bentivoglio M (2000a) Co-induction of nitric oxide synthase, bcl-2 and growth-associated protein-43 in spinal motoneurons during axon regeneration in the lizard tail. Neuroscience 101:451–458
Cristino L, Pica A, Corte FD, Bentivoglio M (2000b) Plastic changes and nitric oxide synthase induction in neurons that innervate the regenerated tail of the lizard Gekko gecko. I. Response of spinal mononeurons to tail amputation and regeneration. J Comp Neurol 417:60–72

Cristino L, Florenzano F, Bentivoglio M, Guglielmotti V (2004) Nitric oxide synthase expression and cell changes in dorsal root ganglia and spinal dorsal horn of developing and adult Rana esculenta indicate a role of nitric oxide in limb metamorphosis. J Comp Neurol 472:423–436

Crossley DA, Wang T, Altimiras J (2000) Role of nitric oxide in the systemic and pulmonary circulation of anesthetized turtles (Trachemys scripta). J Exp Zool 286:683–689

Cuenca N, Naverkamp S, Kolb H (2000) Choline acetyltransferase is found in terminals of horizontal cells that label with $GABA_A$, nitric oxide synthase and calcium binding proteins in the turtle retina. Brain Res 878:228–239

Cutajar MC, Edwards TM, Ng KT (2005) Inhibition of endogenous carbon monoxide production induces transient retention losses in the day-old chick when trained using a single trial passive avoidance learning task. Neurobiol Learn Mem 83:243–250

Cuthbertson S, Jackson B, Toledo C, Fitzgerald ME, Shih YF, Zagvazdin Y, Reiner A (1997) Innervation of orbital and choroidal blood vessels by the pterygopalatine ganglion in pigeons. J Comp Neurol 386:422–442

Cuthbertson S, Zagvazdin YS, Kimble TD, Lamoreaux WJ, Jackson BS, Fitzgerald ME, Reiner A (1999) Preganglionic endings from nucleus of Edinger-Westphal in pigeon ciliary ganglion contain neuronal nitric oxide synthase. Vis Neurosci 16:819–834

Dahlem YA, Hanke W (2005) Intrinsic optical signal of retinal spreading depression: second phase depends on energy metabolism and nitric oxide. Brain Res 1049:15–24

Derer P, Derer M (1993) Ontogenesis of NADPH-diaphorase neurons in the mouse forebrain. Neurosci Lett 152:21–24

Descarries LM, Cai S, Robitaille R, Josephson EM, Morest DK (1998) Localization and characterization of nitric oxide synthase at the frog neuromuscular junction. J Neurocytol 27:829–840

Dimmeler S, Fleming I, Fisslthaler B, Hermann C, Busse R, Zeiher AM (1999) Activation of nitric oxide synthase in endothelial cells by Akt-dependent phosphorylation. Nature 399:601–605

Donald JA, Broughton BR (2005) Nitric oxide control of lower vertebrate blood vessels by vasomotor nerves. Comp Biochem Physiol A Mol Integr Physiol 142:188–197

Donald JA, Broughton BR, Bennett MB (2004) Vasodilator mechanisms in the dorsal aorta of the giant shovelnose ray, Rhinobatus typus (Rajiformes; Rhinobatidae). Comp Biochem Physiol A Mol Integr Physiol 137:21–31

Doran SA, Tran CH, Eskicioglu C, Stachniak T, Ahn KC, Goldberg JI (2003) Constitutive and permissive roles of nitric oxide activity in embryonic ciliary cells. Am J Physiol 285:R348–R355

Duan Y, Panoff J, Burrell BD, Sahley CL, Muller KJ (2005) Repair and regeneration of functional synaptic connections: cellular and molecular interactions in the leech. Cell Mol Neurobiol 25:441–450

Ebbesson LO, Tipsmark CK, Holmqvist B, Nilsen T, Andersson E, Stefansson SO, Madsen SS (2005) Nitric oxide synthase in the gill of Atlantic salmon: colocalization with and inhibition of Na+,K+-ATPase. J Exp Biol 208:1011–1017

Eddy FB, Tibbs P (2003) Effects of nitric oxide synthase inhibitors and a substrate, L-arginine, on the cardiac function of juvenile salmonid fish. Comp Biochem Physiol C Toxicol Pharmacol 135:137–144

Edwards TM, Rickard NS (2002) Inhibition of mono ADP-ribosylation prevents long-term memory consolidation of a single-trial passive avoidance task in the day-old chick. Neurobiol Learn Mem 78:192–198

Edwards TM, Rickard NS (2005) A scavenger of peroxynitrite prevents long-term memory formation using a single trial passive avoidance task for the day-old chick. Neurobiol Learn Mem 83:163–167

Edwards TM, Rickard NS, Ng KT (2002) Inhibition of guanylate cyclase and protein kinase G impairs retention for the passive avoidance task in the day-old chick. Neurobiol Learn Mem 77:313–326

El Dwairi Q, Guo Y, Comtois A, Zh E, Greenwood MT, Bredt DS, Hussain SN (1998) Ontogenesis of nitric oxide synthases in the ventilatory muscles. Am J Respir Cell Mol Biol 18:844–852

Esteban FJ, Jimenez A, Barroso JB, Pedrosa JA, Del Moral MI, Rodrigo J, Peinado MA (1998) The innervation of rainbow trout (Oncorhynchus mykiss) liver: protein gene product 9.5 and neuronal nitric oxide synthase immunoreactivities. J Anat 193:241–249

Evans DH (2001) Vasoactive receptors in abdominal blood vessels of the dogfish shark, Squalus acanthias. Physiol Biochem Zool 74:120–126

Evans DH, Gunderson MP (1998) A prostaglandin, not NO, mediates endothelium-dependent dilation in ventral aorta of shark (Squalus acanthias). Am J Physiol 274:R1050–R1057

Evans DH, Harrie AC (2001) Vasoactivity of the ventral aorta of the American eel (Anguilla rostrata), Atlantic hagfish (Myxine glutinosa), and sea lamprey (Petromyzon Marinus). J Exp Zool 289:273–284

Fatehi-Hassanabad Z, Fatehi M (2004) Characterization of some pharmacological effects of the venom from Vipera lebetina. Toxicon 43:385–391

Fenoglio C, Visai L, Addario C, Gerzeli G, Milanesi G, Vaccrone R, Barni S (2004) Expression of natriuretic peptides, nitric oxide synthase, and guanylate cyclase activity in frog mesonephros during the annual cycle. Gen Comp Endocrinol 137:166–176

Fischer AJ, Stell WK (1999) Nitric oxide synthase-containing cells in he retina, pigmented epithelium, choroid, and sclera of the chick eye. J Comp Neurol 405:1–14

Fleming I, Fissithaler B, Dimmeler S, Kemp BE, Busse R (2001) Phosphorylation of Th495 regulates Ca^{2+}/calmodulin-dependent endothelial nitric oxide synthase activity. Circ Res 88:e68–e75

Fock EM, Lavrova EA, Bachteeva VT, Chernigovskaya EV, Parnova RG (2004) Nitric oxide inhibits arginine-vasotocin-induced increase of water osmotic permeability in frog urinary bladder. Pflugers Arch 448:197–203

Förstermann U, Pollock JS, Schmidt HH, Heller M, Murad F (1991) Calmodulin-dependent endothelium-derived relaxing factor/nitric oxide synthase activity is present in the particulate and cytosolic fractions of bovine aortic endothelial cells. Proc Natl Acad Sci USA 88:1788–1792

Fritsche R, Schwerte T, Pelster B (2000) Nitric oxide and vascular reactivity in developing zebrafish, Danio rerio. Am J Physiol 279:R2200–R2207

Fujii S, Honda S, Sekiya Y, Yamasaki M, Yamamoto M, Saijoh K (1998) Differential expression of nitric oxide synthase isoforms in form-deprived chick eyes. Curr Eye Res 17:586–593

Fujikado T, Kawasaki Y, Fujii J, Taniguchi N, Okada M, Suzuki A, Ohmi G, Tano Y (1997) The effect of nitric oxide synthase inhibitor on form-deprivation myopia. Curr Eye Res 16:992–996

Fujikado T, Tsujikawa K, Tamura M, Hosohata J, Kaweasaki Y, Tano Y (2001) Effect of nitric oxide synthase inhibitor on lens-induced myopia. Ophthalmic Res 33:75–79

Fulton D, Gratton JP, McCabe TJ, Fontana J, Fujio Y, Walsh K, Franke TF, Papapetropoulos A, Sessa WC (1999) Regulation of endothelium-derived nitric oxide production by the protein kinase Akt. Nature 399:597–601

Funakoshi K, Kadota T, Atobe Y, Nakano M, Goris RC, Kishida R (1999) Nitric oxide synthase in the glossopharyngeal and vagal afferent pathway of a teleost, Takifugu niphobles. The branchial vascular innervation. Cell Tissue Res 298:45–54

Furchgott RF (1988) Studies on relaxation of rabbit aorta by sodium nitrite: the basis for the proposal that the acid-activatable inhibitory factor from bovine retractor penis is inorganic nitrite and the endothelium-derived relaxing factor is nitric oxide. In: Vanhoutte PM (ed) Vasodilatation: vascular smooth muscle, peptide, autonomic nerves, and endothelium. Raven Press, New York, pp 401–414

Furchgott RF, Zawadzki JV (1980) The obligatory role of endothelial cells in the relaxation of arterial smooth muscle by acetylcholine. Nature 288:373–376

Galli GL, Skovgaard N, Abe AS, Taylor EW, Wang T (2005) The role of nitric oxide in the regulation of the systemic and pulmonary vasculature of the rattlesnake, Crotalus durissus terrificus. J Comp Physiol [B] 175:201–208

Gallo VP, Civinini A (2001) Immunohistochemical localization of nNOS in the head kidney of larval and juvenile rainbow trout Oncorhynchus mykiss. Gen Comp Endocrinol 124:21–29

Garcia-Cardena G, Oh P, Liu J, Scnitzer JE, Sessa WC (1996) Targeting of nitric oxide synthase to endothelia cell caveolae via palmitoylation: implications for nitric oxide signaling. Proc Natl Acad Sci USA 93:6448–6453

Gardino PF, Schmal AR, Calaza Kda C (2004) Identification of neurons with acetylcholinesterase and NADPH-diaphorase activities in the centrifugal visual system of the chick. J Chem Neuroanat 27:267–273

Garthwaite J, Southam E, Boulton CL, Nielsen EB, Schmidt K, Mayer B (1985) Potential and selective inhibition of nitric oxide-sensitive guanylate cyclase by 1H[1,2,4]oxadiazolo[4,3-a]quinoxalin-1-on. Mol Pharmacol 48:184–188

Genestra M, de Souza WJ, Cysne-Finkelstein L, Leon LL (2003a) Comparative analysis of the nitric oxide production by Leishmania sp. Med Microbiol Immunol (Berl) 192:217–223

Genestra M, Cysne-Finkelstein L, Guedes-Silva D, Leon LL (2003b) Effect of L-arginine analogs and a calcium chelator on nitric oxide (NO) production by Leishmania sp. J Enzyme Inhib Med Chem 18:445–452

Gibbins IL, Olsson C, Holmgren S (1995) Distribution of neurons reactive for NADPF-diaphorase in the branchial nerves of a teleost fish, Gadus morhua. Neurosci Lett 193:113–116

Gossrau R (1998) Caveolin-3 and nitric oxide synthase I in healthy and diseased skeletal muscle. Acta Histochem 100:99–112

Graves AR, Lewin KA, Lindgren CA (2004) Nitric oxide, cAMP and the biphasic muscarinic modulation of ACh release at the lizard neuromuscular junction. J Physiol 559:423–432

Green K, Campbell G (1994) Nitric oxide formation is involved in vagal inhibition of the stomach of the trout (Salmo gairdneri). J Auton Nerv Syst 50:221–229

Greenfield MD (2001) Missing link in firefly bioluminescence revealed: NO regulation of photocyte respiration. Bioessays 23:992–995

Griffith OW, Stuehr DJ (1995) Nitric oxide synthases: properties and catalytic mechanism. Annu Rev Physiol 57:707–736

Grody WW, Dizikes GJ, Cederbaum SD (1987) Human arginase isozymes. Isozymes Curr Top Biol Med Res 13:181–214

Groenendijk BC, Hierck BP, Gittenberger-De Groot AC, Poelmann RE (2004) Development-related changes in the expression of shear stress responsive genes KLF-2, ET-1, and NOS-3 in the developing cardiovascular system of chicken embryos. Dev Dyn 230:57–68

Groenendijk BC, Hierck BP, Vrolijk J, Baiker M, Pourquie MJ, Gittenberger-de Groot AC, Poelmann RE (2005) Changes in shear stress-related gene expression after experimentally altered venous return in the chicken embryo. Circ Res 96:1291–1298

Gruetter CA, Gruetter DY, Lyon EJ, Kadowitz PJ, Ignarro LJ (1981) Relationship between cyclic guanosine 3′,5′-monophosphate formation and relaxation of coronary artery smooth muscle by glyceryltrinitrate, nitroprusside, nitrite, and nitric oxide: effects of methylene blue and methemoglobin. J Pharmacol Exp Ther 219:181–186

Guirado S, Real MA, Olmos JL, Davila JC (2003) Distinct types of nitric oxide-producing neurons in the developing and adult mouse claustrum. J Comp Neurol 465:431–444

Gulati P, Chan AS, Leong SK (1993) NADPH-diaphorase positive cells in the chick and rat thymus. Thymus 22:117–124

Gulati P, Chan AS, Leong SK (1995) Ultrastructural localisation of NADPH-diaphorase in the chick thymic medulla. Cell Tissue Res 279:4-5–409

Gulati P, Tay SS, Leong SK (1997) Nitrergic, peptidergic and substance P innervation of the chick thymus. J Hirnforsch 38:553–564

Gulati P, Leong S, Chan AS (1998) Ontogenesis of NADPH-d expression in the thymic microenvironment of the chick embryo. Cell Tissue Res 294:335–343

Gwee MC, Gopalakrishnakone P, Cheah LS, Bowman WC (1995) L-arginine-nitric oxide pathway involvement in the nerve-evoked relaxant responses of the 5-HT precontracted chick isolated upper oesophagus. J Auton Pharmacol 15:151–158

Hagiwara S, Nakajima S (1966) Differences in Na and Ca spikes as examined by application of tetrodotoxin, procaine, and manganese ions. J Gen Physiol 49:793–806

Hallemeesch MM, Lamers WH, Deutz NE (2002) Reduced arginine availability and nitric oxide production. Clin Nutr 21:273–279

Halm MP, Chichery MP, Chichery R (2003) Effect of nitric oxide synthase inhibition on the manipulative behaviour of Sepia officinalis. Comp Biochem Physiol C Toxicol Pharmacol 134:139–146

Haraldsen L, Soderstrom-Lauritzsen V, Nilsson GE (2002) Oxytocin stimulates cerebral blood flow in rainbow trout (Oncorhynchus mykiss) through a nitric oxide dependent mechanism. Brain Res 929:10–14

Harris MB, Ju H, Venema VJ, Liang H, Zou R, Michell BJ, Chen ZP, Kemp BE, Venema RC (2001) Reciprocal phosphorylation and regulation of endothelial nitric-oxide synthase in response to bradykinin stimulation. J Biol Chem 276:16587–16591

Hasegawa K, Nishimura H (1991) Humoral factor mediates acetylcholine-induced endothelium-dependent relaxation of chicken aorta. Gen Comp Endocrinol 84:164–169

Haverkamp S, Eldred WD (1998a) Localization of nNOS in photoreceptor, bipolar and horizontal cells in turtle and rat retinas. Neuroreport 9:2231–2235

Haverkamp S, Eldred WD (1998b) Localization of the origin of retinal efferents in the turtle brain and the involvement of nitric oxide synthase. J Comp Neurol 393:185–195

Haverkamp S, Kolb H, Cuenca N (1999) Endothelial nitric oxide (eNOS) is localized to Muller cells in all vertebrate retinas. Vision Res 39:2299–2303

Haverkamp S, Kolb H, Cuenca N (2000) Morphological and neurochemical diversity of neuronal nitric oxide synthase-positive amacrine cells in the turtle retina. Cell Tissue Res 302:11–19

Hedrick MS, Morales RD (1999) Nitric oxide as a modulator of central respiratory rhythm in the isolated brainstem of the bullfrog (Rana catesbeiana). Comp Biochem Physiol A Mol Integr Physiol 124:243–251

Hedrick MS, Chen AK, Jessop KL (2005) Nitric oxide changes its role as a modulator of respiratory motor activity during development in the bullfrog (Rana catesbeiana). Comp Biochem Physiol A Mol Integr Physiol 142:231–240

Heinrich UR, Lioudyno M, Maurer J, Mann W, Guth PS, Forstermann U (2003) Localization of the two constitutively expressed nitric oxide synthase isoforms (nNOS and eNOS) in the same cell types in the saccule maculae of the frog Rana pipiens by immunoelectron microscopy: evidence for a back-up system? J Electron Microsc (Tokyo) 52:197–206

Hevel JM, White KA, Marletta MA (1991) Purification of the inducible murine macrophage nitric oxide synthase: identification as a flavoprotein. J Biol Chem 266:22789–22791

Hiramatsu K, Ohshima K (1994) Colocalization of NADPH-diaphorase with neuropeptides in the intrapancreatic neurons of the chicken. Neurosci Lett 182:37–40

Hiramatsu K, Kawamori Y, Ohshima K (1999) Histochemical study of the distribution of nicotinamide adenine dinucleotide phosphate diaphorase (NADPHd)-positive neurons in the chicken caecum. Anat Histol Embryol 28:345–349

Ho KH, Kwan CY, Huang SJ, Bourreau JP (1998) Dual effects of cobra cardiotoxin on vascular smooth muscle and endothelium. Zhongguo Yao Li Xue Bao 19:197–202

Hohler B, Mayer B, Kummer W (1994) Nitric oxide synthase in the rat carotid body and carotid sinus. Cell Tissue Res 276:559–564

Holdengreber V, Krieger C, Gutlich M, Schramek N, Vechoropoulos M, Fischer M, Bacher A, Ben-Shaul Y (2002) Localization of two enzymes of the tetrahydrobiopterin biosynthetic pathway in embryonic chick retina. J Histochem Cytochem 50:265–274

Holmberg A, Kaim J, Persson A, Jensen J, Wang T, Holmgren S (2002) Effects of digestive status on the reptilian gut. Comp Biochem Physiol A Mol Integr Physiol 133:499–518

Holmqvist B, Ellingsen B, Alm P, Forsell J, Oyan AM, Goksoyr A, Fjose A, Seo HC (2000) Identification and distribution of nitric oxide synthase in the brain of adult zebrafish. Neurosci Lett 292:119–122

Holmqvist BI, Ostholm T, Alm P, Ekstrom P (1994) Nitric oxide synthase in the brain of a teleost. Neurosci Lett 171:205–208

Holscher C (1994) 7-Nitro indazole, a neuron-specific nitric oxide synthase inhibitor, produces amnesia in the chick. Learn Mem 1:213–216

Holscher C, Rose SP (1992) An inhibitor of nitric oxide synthesis prevents memory formation in the chick. Neurosci Lett 145:165–167

Holscher C, Rose SP (1993) Inhibiting synthesis of the putative retrograde messenger nitric oxide results in amnesia in a passive avoidance task in the chick. Brain Res 619:189–194

Hopkins BJ, Hodgson WC (1998) Cardiovascular studies on venom from the soldierfish (Gymnapistes marmoratus). Toxicon 36:973–983

Hu J, Li S, Xiao Q, Wang SR (2001) Tecto-isthmo-optic transmission in pigeons is mediated by glutamate and nitric oxide. Brain Res Bull 54:399–403

Hylland P, Nilsson GE (1995) Evidence that acetylcholine mediates increased cerebral blood flow velocity in crucian carp through a nitric oxide-dependent mechanism. J Cereb Blood Flow Metab 15:519–524

Hylland P, Nilsson GE, Lutz PL (1996) Role of nitric oxide in the elevation of cerebral blood flow induced by acetylcholine and anoxia in the turtle. J Cereb Blood Flow Metab 16:290–295

Iadecola C (1992) Does nitric oxide mediate the increases in cerebral blood flow elicited by hypercapnia? Proc Natl Acad Sci USA 89:3913–3916

Ie Noble FA, Ruijtenbeek K, Gommers S, de Mey JG, Blanco CE (2000) Contractile and relaxing reactivity in carotid and femoral arteries of chicken embryos. Am J Physiol 278:H1261–H1268

Ignarro LJ, Buga GM, Wood KS, Byrns RE, Chaudhuri G (1987) Endothelium-derived relaxing factor produced and released from artery and vein is nitric oxide. Proc Natl Acad Sci USA 84:9265–9269

Ignarro LJ, Bush PA, Buga GM, Wood KS, Fukuto JM, Rajfer J (1990) Nitric oxide and cyclic GMP formation upon electrical field stimulation cause relaxation of corpus cavernosum smooth muscle. Biochem Biophys Res Commun 170:843–850

Imbrogno S, De Iuri L, Mazza R, Tota B (2001) Nitric oxide modulates cardiac performance in the heart of Anguilla anguilla. J Exp Biol 204:1719–1727

Jennings BL, Broughton BR, Donald JA (2004) Nitric oxide control of the dorsal aorta and the intestinal vein of the Australian short-finned eel Anguilla australis. J Exp Biol 207:1295–1303

Jewett DC, Butelman ER, Woods JH (1996) Nitric oxide synthase inhibitors produce phencyclidine-like behavioral effects in pigeons. Brain Res 715:25–31

Jimenez A, Esteban FJ, Sanchez-Lopez AM, Pedrosa JA, Del Moral ML, Hernandez R, Blanco S, Barroso JB, Rodfigo J, Peinado MA (2001) Immunohistochemical localization of neuronal nitric oxide synthase in the rainbow trout kidney. J Chem Neuroanat 21:289–294

Ju H, Zou R, Venema VJ, Venema RC (1997) Direct interaction of endothelial nitric oxide synthase and caveolin-1 inhibits synthase activity. J Biol Chem 272:18522–18525

Kagstrom J, Holmgren S (1997) VIP-induced relaxation of small arteries of the rainbow trout, Oncorhynchus mykiss, involves prostaglandin synthesis but not nitric oxide. J Auton Nerv Syst 63:68–76

Kao CY (1972) Pharmacology of tetrodotoxin and saxitoxin. Fed Proc 31:1117–1123

Katsuki S, Arnold W, Mittal C, Murad F (1977) Stimulation of guanylate cyclase by sodium nitroprusside, nitroglycerin and nitric oxide in various tissue preparations and comparison to the effects of sodium azide and hydroxylamine. J Cyclic Nucleotide Res 3:23–35

Kerschbaum HH, Huang S, Xie M, Hermann A (2000) NADPH-diaphorase activity and nitric oxide synthase activity in the kidney of the clawed frog, Xenopus laevis. Cell Tissue Res 301:405–411

Kitazawa T, Onodera C, Taneike T (2002) Potentiation of motilin-induced contraction by nitric oxide synthase inhibition in the isolated chicken gastrointestinal tract. Neurogastroenterol Motil 14:3–13

Kline LW, Pang PK (1994) Calcitonin gene-related peptide relaxes cholecystokinin-induced tension in Iguana iguana gallbladder strips. Gen Comp Endocrinol 95:381–386

Knight GE, Burnstock G (1996) The involvement of the endothelium in the relaxation of the leopard frog (Rana pipiens) aorta in response to acetylcholine. Br J Pharmacol 118:1518–1522

Knight GE, Burnstock G (1999) NANC relaxation of the circular smooth muscle of the oesophagus of the Agama lizard involves the L-arginine-nitric oxide synthase pathway. Comp Biochem Physiol C Pharmacol Toxicol Endocrinol 122:165–171

Kohn AB, Moroz LL, Lea JM, Greenberg RM (2001) Distribution of nitric oxide synthase immunoreactivity in the nervous system and peripheral tissues of Schistosoma mansoni. Parasitology 122:87–92

Kronstrom J, Holmgren S, Baguet F, Salpietro L, Mallefet J (2005) Nitric oxide in control of luminescence from hatchetfish (Argyropelecus hemigymnus) photophores. J Exp Biol 208:2951–2961

Kurylas AE, Ott SR, Schachtner J, Elphick MR, Williams L, Homberg U (2005) Localization of nitric oxide synthase in the central complex and surrounding midbrain neuropils of the locust Schistocerca gregaria. J Comp Neurol 484:206–223

Kwon NS, Nathan DF, Stuehr DJ (1989) Reduced biopterin as cofactor in the generation of nitrogen oxide by murine macrophages. J Biol Chem 264:20496–20501

Lamanna C, Costagliola A, Vittoria A, Mayer B, Assisi L, Botte V, Cecio A (1999) NADPH-diaphorase and NOS enzymatic activities in some neurons of reptilian gut and their relationship with two neuropeptides. Anat Embryol (Berl) 199:397–405

Lamanna C, Vittoria A, Assisi L, Lucini C, Castaldo L, Botte V (2001) NADPH-d positive neurons of the lizard Podarcis s. sicula oviduct and their relationship to 17β-estradiol hormone. Life Sci 69:1765–1773

Lee CY, Zou HS, Yau SM, Ju YR, Liau CS (2000) Nitric oxide synthase activity and immunoreactivity in the crayfish Procambarus clarkii. Neuroreport 11:1273–1276

Levy H, Twig G, Perlman I (2004) Nitric oxide modulates the transfer function between cones and horizontal cells during changing conditions of ambient illumination. Eur J Neurosci 20:2963–2974

Li ZS, Furness JB (1993) Nitric oxide synthase in the enteric nervous system of the rainbow trout, Salmo gairdneri. Arch Histol Cytol 56:185–193

Liepe BA, Stone C, Koistinaho J, Copenhagen DR (1994) Nitric oxide synthase in Muller cells and neurons of salamander and fish retina. J Neurosci 14:7641–7654

Lin YQ, Bennett MR (1994) Nitric oxide modulation of quantal secretion in chick ciliary ganglia. J Physiol 481:385–394

Low KS, Gwee MC, Yuen R, Gopalakrishnakone P, Khoo HE (1993) Stonustoxin: a highly potent endothelium-dependent vasorelaxant in the rat. Toxicon 31:1471–1478

M'Rabet-Touil H, Blachier F, Morel MT, Darcy-Vrillon B, Duee PH (1993) Characterization and ontogenesis of nitric oxide synthase activity in pig enterocytes. FEBS Lett 331:243–247

Ma PM (1993) Tanycytes in the sunfish brain: NDPH-diaphorase histochemistry and regional distribution. J Comp Neurol 336:77–95

Maake C, Kaufmann C, Reinecke M (2001) Ontogeny of neurohumoral peptides, serotonin, and nitric oxide synthase in the gastrointestinal neuroendocrine system of the axolotl (Ambystoma mexicanum): an immunohistochemical analysis. Gen Comp Endocrinol 121:74–83

Manwar SJ, Moudgal RP, sastry KV, Mohan J, Tyagi JB, Raina R (2006) Role of nitric oxide in ovarian follicular development and egg production in Japanese quail (Coturnix coturnix japonica). Theriogenology 65:1392–1400

Martin MT, Gonalons E, Fernandez E (1998) Contribution of inhibitory neurotransmission to the CCK induced relaxation of the circular muscle of avian ileum. Life Sci 62:937–946

Martinez V, Jimenez M, Gonalons E, Vergara P (1993) Mechanism of action of CCK in avian gastroduodenal motility: evidence for nitric oxide involvement. Am J Physiol 265:G842–G850

Martinez-Ciriano C, Junquera C, Castiella T, Gomez-Barrena E (2000) Intrinsic innervation in the intestine of the lizard Podarcis hispanica. Histol Histopathol 15:1093–1105

Martinez-Lemus LA, Hester RK, Becker EJ, Jeffrey JS, Odom TW (1999) Pulmonary artery endothelium-dependent vasodilation is impaired in a chicken model of pulmonary hypertension. Am J Physiol 277:R190–R197

Maury P, Sarre A, Terrand J, Rosa A, Kucera P, Kappenberger L, Raddatz E (2004) Ventricular but not atrial electro-mechanical delay of the embryonic heart is altered by anoxia-reoxygenation and improved by nitric oxide. Mol Cell Biochem 265:141–149

McNeill B, Perry SF (2005) Nitric oxide and the control of catecholamine secretion in rainbow trout Oncorhynchus mykiss. J Exp Biol 208:2421–2431

Meng W, Ayata C, Waeber C, Juang PL, Mozkowitz MA (1998) Neuronal NOS-cGMP-dependent ACh-induced relaxation in pial arterioles of endothelial NOS knockout mice. Am J Physiol 274:H411–H415

Meyer G, Banuelos-Pineda J, Montagnese C, Ferres-Meyer G, Gonzalez-Hernandez T (1994) Laminar distribution and morphology of NADPH-diaphorase containing neurons in the optic tectum of the pigeon. J Hirnforsch 35:445–452

Michel JB, Feron O, Sacks D, Michel T (1997) Reciprocal regulation of endothelial nitric-oxide synthase by Ca^{2+}-calmodulin and caveolin. J Biol Chem 272:15583–15586

Miller VM, Vanhoutte PM (2000) Prostaglandins but not nitric oxide are endothelium-derived relaxing factors in the trout aorta. Acta Pharmacol Sin 21:871–876

Mirabella N, Germano G, Langella M, Lamanna C, Paino G (2000a) The distribution and colocalization of nicotinamide adenine dinucleotide phosphate reduced-diaphorase (NADPH-d)-/nitric oxide synthase (NOS)-containing neurons in the innervation of the duck ureter. Anat Embryol (Berl) 202:291–301

Mirabella N, Lamanna C, Assisi L, Botte V, Cecio A (2000b) The relationships of nicotinamide adenine dinucleotide phosphate-d to nitric oxide synthase, vasoactive intestinal polypeptide, galanin and pituitary adenylate activating polypeptide in pigeon gut neurons. Neurosci Lett 293:147–151

Miyabara EH, Tostes RC, Selistre-de-Araujo HS, Aoki MS, Moriscot AS (2004) Role of nitric oxide in myotoxic activity induced by crotoxin in vivo. Toxicon 43:425–432

Miyachi E, Murakami M, Nakaki T (1990) Arginine blocks gap junctions between retinal horizontal cells. Neuroreport 1:107–110

Moncada S, Radomski MW, Palmer RM (1988) Endothelium-derived relaxing factor: identification as nitric oxide and role in the control of vascular tone and platelet function. Biochem Pharmacol 37:2495–2501

Moon C, Terashima S, Ahn M, Kang J, Shin T (2002) Immunohistochemical analysis of neuronal nitric oxide synthase in the trigeminal ganglia of the crotaline snake Trimeresurus flavoviridis. Neurosci Lett 319:21–24

Moon C, Terashima S, Shin T (2004) Immunohistochemical study of endothelial nitric oxide synthase in the trigeminal ganglia of a crotaline snake Trimeresurus flavoviridis. J Vet Med Sci 66:1007–1009

Moore PK, Babbedge RC, Wallace P, Gaffen ZA, Hart SL (1993) 7-Nitroindazole, an inhibitor of nitric oxide synthase, exhibits anti-nociceptive activity in the mouse without increasing blood pressure. Br J Pharmacol 108:296–297

Moreno de Sandino M, Hernandez A (2003) Nitric oxide synthase expression in the endothelium of pulmonary arterioles in normal and pulmonary hypertensive chickens subjected to chronic hypobaric hypoxia. Avian Dis 47:1291–1297

Moreno N, Lopez JM, Sanchez-Camacho C, Gonzalez A (2002a) Development of NADPH-diaphorase/nitric oxide synthase in the brain of the urodele amphibian Pleurodeles waltl. J Chem Neuroanat 23:105–121

Moreno N, Lopez JM, Sanchez-Camacho C, Crespo M, Munoz M, Gonzalez A (2002b) Early development of NADPH diaphorase-expressing neurons in the brain of the urodele amphibian Pleurodeles waltl. Brain Res Bull 57:409–412

Morgan IG, Miethke P, Li ZK (1994) Is nitric oxide a transmitter of the centrifugal projection to the avian retina? Neurosci Lett 168:5–7

Moroz LL, Chen D, Gillette MU, Gillette R (1996) Nitric oxide synthase activity in the molluscan CNS. J Neurochem 66:873–876

Moroz LL, Dhlgren RL, Boudko D, Sweedler JV, Lovell P (2005) Direct single cell determination of nitric oxide synthase related metabolites in identified nitrergic neurons. J Inorg Biochem 99:929–939

Moura JR, Prado MA, Gomez MV, Kalapothakis E, Diniz CR, Cordeiro MN, Romano-Silva MA (1998) Investigation of the effect of PhTx 2, from the venom of the spider Phoneutria nigriventer, on the release of [3H]-acetylcholine from rat cerebrocortical synaptosomes. Toxicon 36:1189–1192

Mustafa T, Agnisola C (1998) Vasoactivity of adenosine in the trout (Oncorhynchus mykiss) coronary system: involvement of nitric oxide and interaction with noradrenaline. J Exp Biol 201:3075–3083

Nagayama T, Hosokawa A, Yoshida M, Suzuki-Kusaba M, Hisa H, Kimura T, Satoh S (1998) Role of nitric oxide in adrenal catecholamine secretion in anesthetized dogs. Am J Physiol 275:R1075–R1081

Nakane M, Schmidt HH, Pollock JS, Forstermann U, Murad F (1993) Cloned human brain nitric oxide synthase is highly expressed in skeletal muscle. FEBS Lett 316:175–180

Narahashi T, Moore JW, Scott WR (1964) Tetrodotoxin blockage of sodium conductance increase in lobster giant axons. J Gen Physiol 47:965–974

Necker R (2004) Distribution of choline acetyltransferase and NADPH diaphorase in the spinal cord of the pigeon. Anat Embryol (Berl) 208:169–181

Necker R (2005) Embryonic development of choline acetyltransferase and nitric oxide synthase in the spinal cord of pigeons and chickens with special reference to the superficial dorsal horn. Anat Embryol (Berl) 210:145–154

Nichol KA, Chan N, Davey DF, Bennett MR (1995) Location of nitric oxide synthase in the developing avian ciliary ganglion. J Auton Nerv Syst 51:91–102

Nickla DL, Wildsoet CF (2004) The effect of the nonspecific nitric oxide synthase inhibitor NG-nitro-L-arginine methyl ester on the choroidal compensatory response to myopic defocus in chickens. Optom Vis Sci 81:111–118

Nilsson GE, Soderstrom V (1997) Comparative aspects on nitric oxide in brain and its role as a cerebral vasodilator. Comp Biochem Physiol 118:949–958

Ny L, Person K, Larsson B, Chan J, Weiss LM, Wittner M, Huang H, Tanowitz HB (1999) Localization and activity of nitric oxide synthases in the gastrointestinal tract of Trypanosoma cruzi-infected mice. J Neuroimmunol 99:27–35

Odom TW, Martinez-Lemus LA, Hester RK, Becker FJ, Jeffrey JS, Meininger GA, Ramirez GA (2004) In vivo hypoxia differentially affects constriction and relaxation responses of isolated pulmonary arteries from broiler and leghorn chickens. Poult Sci 83:835–841

Olgart C, Iversen HH (1999) Nitric oxide-dependent relaxation induced by M1 muscarinic receptor activation in the rat small intestine. Br J Pharmacol 127:309–313

Olmos JL, Real MA, Medina L, Guirado S, Davila JC (2005) Distribution of nitric oxide-producing neurons in the developing and adult muse amygdalar basolateral complex. Brain Res Bull 66:465–469

Olson KR, Villa J (1991) Evidence against nonprostanoid endothelium-derived relaxing factor(s) in trout vessels. Am J Physiol 260:R925–R933

Olson KR, Conklin DJ, Farrell AP, Keen JE, Takei Y, Weaver L Jr, Smith MP, Zhang Y (1997) Effects of natriuretic peptides and nitroprusside on venous function in trout. Am J Physiol 273:R527–R539

Olsson C, Gibbins I (1999) Nitric oxide synthase in the gastrointestinal tract of the estuarine crocodile, Crocodylus porous. Cell Tissue Res 296:433–437

Oyan AM, Nilsen F, Goksoyr A, Holmqvist B (2000) Partial cloning of constitutive and inducible nitric oxide synthases and detailed neuronal expression of NOS mRNA in the cerebellum and optic tectum of adult Atlantic salmon (Salmo salar). Brain Res Mol Brain Res 78:38–49

Palmer RM, Rees DD, Ashton DS, Moncada S (1988b) L-Arginine is the physiological precursor for the formation of nitric oxide in endothelium-dependent relaxation. Biochem Biophys Res Commun 153:1251–1256

Palmer RMJ, Ferrige AG, Moncada S (1987) Nitric oxide release accounts for the biological activity of endothelium-derived relaxing factor. Nature 327:524–526

Palmer RMJ, Ashton DS, Moncada S (1998a) Vascular endothelial cells synthesize nitric oxide from L-arginine. Nature 333:664–666

Palumbo A (2005) Nitric oxide in marine invertebrates: a comparative perspective. Comp Biochem Physiol A Mol Integr Physiol 142:241–248

Park KH, Kim KH, Choi SH, Yoon JM, Ki YG (2000) Cyclooxygenase-derived products, rather than nitric oxide, are endothelium-derived relaxing factor(s) in the ventral aorta of carp (Cyprinus carpio). Comp Biochem Physiol A Mol Integr Physiol 127:89–98

Partata WA, Krepsky AM, Marques M, Achaval M (1999) Axotomy increases NADPH-diaphorase activity in the dorsal root ganglia and lumbar spinal cord of the turtle Trachemys dorbigni. Braz J Med Biol Res 32:489–493

Pellegrino D, Sprovieri E, Mazza R, Randall DJ, Tota B (2002) Nitric oxide-cGMP-mediated vasoconstriction and effects of acetylcholine in the branchial circulation of the eel. Comp Biochem Physiol A Mol Integr Physiol 132:447–457

Pellegrino D, Tota B, Randall DJ (2005) Adenosine/nitric oxide crosstalk in the branchial circulation of Squalus acanthias and Anguilla anguilla. Comp Biochem Physiol A Mol Integr Physiol 142:198–204

Peunova N, Scheinker V, Cline H, Enikolopov G (2001) Nitric oxide is an essential negative regulator of cell proliferation in Xenopus brain. J Neurosci 21:8809–8818

Picolo G, Cury Y (2004) Peripheral neuronal nitric oxide synthase activity mediates the antinociceptive effect of Crotalus duressus terrificus snake venom, a delta- and kappa-opioid receptor agonist. Life Sci 75:559–573

Picolo G, Giorgi R, Cury Y (2000) Delta-opioid receptors and nitric oxide mediate the analgesic effect of Crotalus duressus terrificus snake venom. Eur J Pharmacol 391:55–62

Pollock JS, Forstermann U, Mitchell JA, Warner TD, Schmidt HH, Nakane M, Murad F (1991) Purification and characterization of particulate endothelium-derived relaxing factor synthase from cultured and native bovine aortic endothelial cells. Proc Natl Acad Sci USA 88:10480–10484

Pollock VP, McGettigan J, Cabrero P, Maudlin IM, Dow JA, Davies SA (2004) Conservation of capa peptide-induced nitric oxide signaling in Diptera. J Exp Biol 207:4135–4145

Postorino A, Vetri T, Bonvissuto F, Fileccia R, Urso SA (1999) Evidence for involvement of nitric oxide (NO) or a related nitroso-compound in NANC inhibitory neurotransmission in the pigeon oesophageal smooth muscle. J Auton Pharmacol 19:85–90

Predel R, Russell WK, Tichy SE, Russell DH, Nachman RJ (2003) Mass spectrometric analysis of putative capa-gene products in Musca domestica and Neobellieria bullata. Peptides 24:1487–1491

Puttmann B, Gerlach EM, Kruger M, Blottner D (2005) Neuromuscular contacts induce nitric oxide signals in skeletal myotubes in vitro. Neurosignals 14:85–95

Rea MS, Parsons RH (2001) Evidence of nitric oxide and angiotensin II regulation of circulation and cutaneous drinking in Bufo marinus. Physiol Biochem Zool 74:127–133

Rees DD, Palmer RMJ, Schulz R, Hodson HF, Moncada S (1990) Characterization of three inhibitors of endothelial nitric oxide synthase in vitro and in vivo. Br J Pharmacol 101:746–752

Renshaw GM, Dyson SE (1999) Increased nitric oxide synthase in the vasculature of the epaulette shark brain following hypoxia. Neuroreport 10:1707–1712

Rickard NS, Gibbs ME (2003a) Hemispheric dissociation of the involvement of NOS isoforms in memory for discriminated avoidance in the chick. Learn Mem 10:314–318

Rickard NS, Gibbs ME (2003b) Effects of nitric oxide inhibition on avoidance learning in the chick are lateralized and localized. Neurobiol Learn Mem 79:252–256

Rickard NS, Ng KT, Gibbs ME (1998) Further support for nitric oxide-dependent memory processing in the day-old chick. Neurobiol Learn Mem 69:79–86

Rickard NS, Gibbs ME, Ng KT (1999) Inhibition of the endothelial isoform of nitric oxide synthase impairs long-term memory formation in the chick. Learn Mem 6:458–466

Rios H, Lopez-Costa JJ, Fosser NS, Brusco A, Saavedra JP (2000) Development of nitric oxide neurons in the chick embryo retina. Brain Res Dev Brain Res 120:17–25

Robertson JD, Bonaventura J, Kohm AP (1994) Nitric oxide is required for tactile learning in Octopus vulgaris. Proc Biol Sci 256:269–273

Rodriguez-Membrilla A, Martinez V, Jimenez M, Gonalons E, Vergara P (1995) Is nitric oxide the final mediator regulating the migrating myoelectric complex cycle? Am J Physiol 268:G207–G214

Ruijtenbeek K, Kessels CG, Villamor E, Blanco CE, De Mey JG (2002) Direct effects of acute hypoxia on the reactivity of peripheral arteries of the chicken embryo. Am J Physiol 283:R331–R338

Ruijtenbeek K, Kessels LC, De Mey JG, Blanco CE (2003) Chronic moderate hypoxia and protein malnutrition both induce growth retardation, but have distinct effects on arterial endothelium-dependent reactivity in the chicken embryo. Pediatr Res 53:573–579

Ruiz-Feria CA, Yang Y, Nishimura H (2004) Do incremental increases in blood pressure elicit neointimal plaques through endothelial injury? Am J Physiol 287:R1486–R1493

Rumbaut RE, McKay MK, Huxley VH (1995) Capillary hydraulic conductivity is decreased by nitric oxide synthase inhibition. Am J Physiol 268:H1856–H1861

Saad WA, Guarda IF, Camargo LA, dos Santos TA, Saad WA, Simoes S, Guarda RS (2002) Novel evidence that nitric oxide of the median septal area influences the Salivary secretion induced by pilocarpine. Life Sci 70:2403–2412

Saad WA, Guarda IF, Camargo LA, dos Santos TA, Guarda RS, Saad WA, Simoes S, Rodrigues JA (2003) Role of nitric oxide of the median preoptic nucleus (MnPO) in the alterations of Salivary flow, arterial pressure and heart rate induced by injection of pilocarpine into the MnPO and intraperitoneally. Braz J Med Biol Res 36:897–905

Sakura M, Kabetani M, Watanabe S, Korino Y (2004) Impairment of olfactory discrimination by blockade of nitric oxide activity in the terrestrial slug Limax valentianus. Neurosci Lett 370:257–261

Samama B, Boehm N (1996) Ontogenesis of NADPH-diaphorase activity in the olfactory bulb of the rat. Brain Res Dev Brain Res 96:192–203

Sanderson NS, Weissler E, Crews D (2005) The nitric oxide synthase inhibitor L-NAME suppresses androgen-induced male-like pseudocopulatory behavior in whiptail lizards. Brain Res 1052:236–239

Sandor P, Komjati K, Reovich M, Nyary I (1994) Major role of nitric oxide in the mediation of regional CO2 responsiveness of the cerebral and spinal cord vessels of the cat. J Cereb Blood Flow Metab 14:49–58

Sann J (1998) Neuronal subpopulations in autonomic ganglia associated with the chicken ureter: an immunohistochemical study. Cell Tissue Res 292:477–485

Savchenko A, Barnes S, Kramer RH (1997) Cyclic-nucleotide-gated channels mediate synaptic feedback by nitric oxide. Nature 390:694–698

Schattner M, Fritzen M, Ventura Jde S, de Albuquerque Modesto JC, Pozner RG, Moura-da-Silva AM, Chudzinski-Tavassi AM (2005) The snake venom metalloproteases berythractivase and jararhagin activate endothelial cells. Biol Chem 386:369–374

Schmidt HH, Pollock JS, Nakane M, Gorsky LD, Forstermann U, Murad F (1991) Purification of a soluble isoform of guanylyl cyclase-activating factor synthase. Proc Natl Acad Sci USA 88:5–369

Schober A, Malz CR, Meyer DL (1993) Enzyme-histochemical demonstration of nitric oxide synthase in the diencephalon of the rainbow trout (Oncorhynchus mykiss). Neurosci Lett 151:67–70

Schrodl F, De Stefano ME, Minvielle F, Brehmer A, Neuhuber WL (2005) Somatostatin immunoreactivity in quail pterygopalatine ganglion. J Anat 206:249–255

Schroedl F, Tines R, Brehmer A, Neuhuber WL (2001) Intrinsic choroidal neurons in the duck eye receive sympathetic input: anatomical evidence for adrenergic modulation of nitrergic functions in the choroid. Cell Tissue Res 304:175–184

Schroedl F, De Stefano ME, Reese S, Brehmer A, Neuhuber WL (2004) Comparative anatomy of nitrergic intrinsic choroidal neurons (ICN) in various avian species. Exp Eye Res 78:187–196

Schwerte T, Axelsson M, Nilsson S, Pelster S (1997) Effects of vagal stimulation on swimbladder blood flow in the European eel Anguilla anguilla. J Exp Biol 200:3133–3139

Schwerte T, Holmgren S, Pelster B (1999) Vasodilation of swimbladder vessels in the European eel (Anguilla anguilla) induced by vasoactive intestinal polypeptide, nitric oxide, adenosine and protons. J Exp Biol 202:1005–1013

Scott TR, Bennett MR (1993) The effect of nitric oxide on the efficacy of synaptic transmission through the chick ciliary ganglion. Br J Pharmacol 110:627–632

Seebacher F, Franklin CE (2004) Integration of autonomic and local mechanisms in regulating cardiovascular responses to heating and cooling in a reptile (Crocodylus porous). J Comp Physiol [B] 174:577–585

Shafer OT, Chen A, Kumar SM, Muller KJ, Sahley CL (1998) Injury-induced expression of endothelial nitric oxide synthase by glial and microglial cells in the leech central nervous system within minutes after injury. Proc Biol Sci 265:2171–2175

Shahbazi F, Conlon JM, Holmgren S, Jensen J (2001) Effects of cod bradykinin and its analogs on vascular and intestinal smooth muscle of the Atlantic cod, Gadus morhua. Peptides 22:1023–1029

Shaul PW, Smart EJ, Robinson LJ, German Z, Yuhanna IS, Ying Y, Anderson RG, Michel T (1996) Acylation targets endothelial nitric oxide synthase to plasmalemmal caveolae. J Biol Chem 271:6518–6522

Shinozaki K, Kashiwagi A, Nishio Y, Okamura T, Yoshida Y, Masada M, Toda N, Kikkawa R (1999) Abnormal biopterin metabolism is a major cause of impaired endothelium-dependent relaxation through nitric oxide/O2– imbalance in insulin-resistance rat aorta. Diabetes 48:2437–2445

Skovgaard N, Galli G, Taylor EW, Conlon JM, Wang T (2005a) Hemodynamic effects of python neuropeptide gamma in the anesthetized python, Python regius. Regul Pept 128:15–26

Skovgaard N, Galli G, Abe A, Taylor EW, Wang T (2005b) The role of nitric oxide in regulation of the Cardiovascular system in reptiles. Comp Biochem Physiol A Mol Integr Physiol 142:205–214

Small SA, MacDonald C, Farrell AP (1990) Vascular reactivity of the coronary artery in rainbow trout (Oncorhynchus mykiss). Am J Physiol 258:R1402–R1410

Soderstrom V, Nilsson GE (2000) Brain blood flow during hypercapnia in fish: no role of nitric oxide. Brain Res 857:207–211

Soderstrom V, Hylland P, Nilsson GE (1995) Nitric oxide synthase inhibitor blocks acetylcholine induced increase in brain blood flow in rainbow trout. Neurosci Lett 197:191–194

Soderstrom V, Nilsson GE, Lutz PL (1997) Effects of inhibition of nitric oxide synthesis and of hypercapnia on blood pressure and brain blood flow in the turtle. J Exp Biol 200:815–820

Soderstrom V, Nilsson GE, Renshaw GM, Franklin CE (1999) Hypoxia stimulates cerebral blood flow in the estuarine crocodile (Crocodylus porous). Neurosci Lett 267:1–4

Staples JF, Zapol WM, Bloch KD, Kawai N, Val VM, Hochachka PW (1995) Nitric oxide responses of air-breathing and water-breathing fish. Am J Physiol 268:R816–R819

Stefano GB, Kim E, Liu Y, Zhu W, Casares F, Manitione K, Jones DA, Cadet P (2004) Nitric oxide modulates microglial activation. Med Sci Monit 10:BR17–BR22

Stuehr DJ, Cho HJ, Kwon NS, Weise MF, Nathan CF (1991a) Purification and characterization of the cytokine-induced macrophage nitric oxide synthase: an FAD- and FMN-containing flavoprotein. Proc Natl Acad Sci USA 88:7773–7777

Stuehr DJ, Kwon NS, Nathan CF, Griffith OW, Feldman PL, Wiseman J (1991b) Nω-hydroxy-arginine is an intermediate in the biosynthesis of nitric oxide from L-arginine. J Biol Chem 266:6259–6263

Sun W, Erichsen JT, May PJ (1994) NADPH-diaphorase reactivity in ciliary ganglion neurons: a comparison of distributions in the pigeon, cat, and monkey. Vis Neurosci 11:1027–1031

Sung JM, Low KS, Khoo HE (2002) Characterization of the mechanism underlying stonustoxin-mediated relaxant response in the rat aorta in vitro. Biochem Pharmacol 63:1113–1118

Sys SU, Pellegrino D, Mazza R, Gattuso A, Andries LJ, Tota L (1997) Endocardial endothelium in the avascular heart of the frog: morphology and role of nitric oxide. J Exp Biol 200:3109–3118

Tan X, Sun WD, Li JC, Pan JQ, Liu YJ, Wang JY, Wang XL (2005a) L-Arginine prevents reduced expression of endothelial nitric oxide synthase (NOS) in pulmonary arterioles of broilers exposed to cool temperatures. Vet J Sep 15; [Epub ahead of print]

Tan X, Pan JQ, Li JC, Liu YJ, Sun WD, Wang XL (2005b) L-Arginine inhibiting pulmonary vascular remodeling is associated with promotion of apoptosis in pulmonary arterioles smooth muscle cells in broilers. Res Vet Sci 79:203–209

Teixeira CE, Teixeira SA, Antunes E, De Nucci G (2001a) The role of nitric oxide on the relaxations of rabbit corpus cavernosum induced by Androctonus Australis and Buthotus judaicus scorpion venom. Toxicon 39:633–639

Teixeira CE, Faro R, Moreno RA, Rodrigues Netto N Jr, Fregonesi A, Antunes E, De Nucci G (2001b) Nonadrenergic, noncholinergic relaxation of human isolated corpus cavernosum induced by scorpion venom. Urology 57:816–820

Teixeira CE, Ifa DR, Corso G, Santagada VA, Caliendo G, Antunes E, De Nucci G (2003) Sequence and structure-activity relationship of a scorpion venom toxin with nitrergic activity in rabbit corpus cavernosum. FASEB J 17:485–487

Terada H, Nagai T, Okada S, Kimura H, Kitahama K (2001) Ontogenesis of neurons immunoreactive for nitric oxide synthase in rat forebrain and midbrain. Brain Res Dev Brain Res 128:121–137

Thomas S, Robitaille R (2001) Differential frequency-dependent regulation of transmitter release by endogenous nitric oxide at the amphibian neuromuscular synapse. J Neurosci 21:1087–1095

Toda N, Herman AG (2005) Gastrointestinal function regulation by nitrergic efferent nerves. Pharmacol Rev 57:315–338

Toda N, Okamura T (1990) Possible role of nitric oxide in transmitting information from vasodilator nerve to cerebral arterial muscle. Biochem Biophys Res Commun 170:308–313

Toda N, Okamura T (2003) The pharmacology of nitric oxide in the peripheral nervous system of blood vessels. Pharmacol Rev 55:271–324

Toda N, Minami Y, Okamura T (1990a) Inhibitory effects of L-NG-nitro-arginine on the synthesis of EDRF and the cerebroarterial response to vasodilator nerve stimulation. Life Sci 47:345–351

Toda N, Baba H, Okamura T (1990b) Role of nitric oxide in non-adrenergic, non-cholinergic nerve-mediated relaxation in dog duodenal longitudinal muscle strips. Jpn J Pharmacol 53:281–284

Toda N, Ayajiki K, Yoshida K, Kimura H, Okamura T (1993) Impairment buy damage of the pterygopalatine ganglion of nitroxidergic vasodilator nerve function in canine cerebral and retinal arteries. Circ Res 72:206–213

Toda N, Kitamura Y, Okamura T (1994) Role of nitroxidergic nerve in dog retinal arterioles in vivo and arteries in vitro. Am J Physiol 266:H1985–H1992

Toda N, Ayajiki K, Okamura T (1997) Inhibition of nitroxidergic nerve function by neurogenic acetylcholine in monkey cerebral arteries. J Physiol 498:453–461

Toda N, Toda M, Ayajiki K, Okamura T (1998) Cholinergic nerve function in monkey ciliary arteries innervated by nitroxidergic nerve. Am J Physiol 274:H1582–H1589

Toda N, Tanaka T, Ayajiki K, Okamura T (2000a) Cerebral vasodilatation induced by stimulation of the pterygopalatine ganglion and greater petrosal nerve in anesthetized monkeys. Neuroscience 96:393–398

Toda N, Ayajiki K, Tanaka T, Okamura T (2000b) Preganglionic and postganglionic neurons responsible for cerebral vasodilatation mediated by nitric oxide in anesthetized dogs. J Cereb Blood Flow Metab 20:700–708

Toda N, Ayajiki K, Okamura T (2005) Nitric oxide and penile erectile function. Pharmacol Ther 106:233–266

Torrao AS, Britto LR (2004) Increased expression of nitric oxide synthase in visual structures of the chick brain after retinal removal. J Neurosci Res 78:123–131

Tota B, Amelio D, Pellegrino D, Ip YK, Cerra MC (2005) NO modulation of myocardial performance in fish hearts. Comp Biochem Physiol A Mol Integr Physiol 142:164–177

Traister A, Abashidze S, Gold V, Plachta N, Karchovsky E, Patel K, Wei M (2002) Evidence that nitric oxide regulates cell-cycle progression in the developing chick neuroepithelium. Dev Dyn 225:271–276

Trimmer BA, Aprille JR, Dudzinski DM, Lagace CJ, Lewis SM, Michel T, Qazi S, Zayas RM (2001) Nitric oxide and the control of firefly flashing. Science 292:2413–2414

Trimmer BA, Aprille J, Modica-Napolitano J (2004) Nitric oxide signalling: insect brains and photocytes. Biochem Soc Symp 71:65–83

Tseng WP, Lin-Shiau SY (2003) Calcium-activated NO production plays a role in neuronal death induced by β-bungarotoxi in primary cultures of cerebellar granular neurons. Naunyn Schmiedebergs Arch Pharmacol 367:451–461

Tucker JF, Brave SR, Charalambopus L, Hobbs AJ, Bibson A (1990) L-NG-nitro arginine inhibits non-adrenergic, non-cholinergic relaxations of guinea-pig isolated tracheal smooth muscle. Br J Pharmacol 100:663–664

Ursell PC, Mayes M (1996) Endothelial isoform of nitric oxide synthase in rat heart increases during development. Anat Rec 246:465–472

Vallance P, Leone A, Calver A, Collier J, Moncada S (1992) Endogenous dimethylarginine as an inhibitor of nitric oxide synthesis. J Cardiovasc Pharmacol 20:S60–S62

Vemulapalli R, Kurowski S, Salisbury B, Parker E, Davis H (2001) Activation of central melanocortin receptors by MT-II increases cavernosal pressure in rabbits by the neuronal release of NO. Br J Pharmacol 134:1705–1710

Verge VM, Xu Z, Xu XJ, Wiesenfeld-Hallin Z, Hokfelt T (1992) Marked increase in nitric oxide synthase mRNA in rat dorsal root ganglia after peripheral axotomy: in situ hybridization and functional studies. Proc Natl Acad Sci USA 89:11617–11621

Vetri T, Postorino A, Fileccia R, Bonvissuto F, Abbadessa Urso S (2004) Involvement of purinergic nerves in the NANC inhibitory junction potentials in pigeon oesophageal smooth muscle. Auton Autacoid Pharmacol 24:1–7

Vicente S, Gonzalez MP, Oset-Gasque MJ (2002) Neuronal nitric oxide synthase modulates basal catecholamine secretion in bovine chromaffin cells. J Neurosci Res 69:327–340

Villamor E, Ruijtenbeek K, Pulgar V, De Mey JG, Blanco CE (2002) Vascular reactivity in intrapulmonary arteries of chicken embryos during transition to ex ovo life. Am J Physiol 282:R917–R927

Villani L (1999) Development of NADPH-diaphorase activity in the central nervous system of the cichlid fish, Tilapia mariae. Brain Behav Evol 54:147–158

Vizzard MA, Erdman SL, de Groat WC (1995) Increased expression of neuronal nitric oxide synthase (NOS) in visceral neurons after nerve injury. J Neurosci 15:4033–4045

von Bartheld CS, Schober A (1997) Nitric oxide synthase in learning-relevant nuclei of the chick brain: morphology, distribution, and relation to transmitter phenotypes. J Comp Neurol 383:135–152

Waldman SA, Murad F (1987) Cyclic GMP synthesis and function. Pharmacol Rev 39:163–196

Wang J, Wang X, Xiang R, Sun W (2002) Effect of L-NAME on pulmonary arterial pressure, plasma nitric oxide and pulmonary hypertension syndrome morbidity in broilers. Br Poult Sci 43:615–620

Weidong S, Xiaolong W, Jinyong W, Ruiping X (2002) Pulmonary arterial pressure and electrocardiograms in broiler chickens infused intravenously with L-NAME, an inhibitor of nitric oxide synthase, or sodium nitroprusside (SNP), a nitric oxide donor. Br Poult Sci 43:306–312

Weinberg M, Cordeiro MN, De L, Oliveira LC, Diniz CR (2002) Endothelium-dependent relaxation of rat mesenteric arterial rings by a Phoneutria nigriventer venom fraction. Jpn J Pharmacol 88:189–196

Wellard JW, Morgan IG (2004) Inhibitory modulation of photoreceptor melatonin synthesis via a nitric oxide-mediated mechanism. Neurochem Int 45:1143–1153

Werner ER, Werner-Felmayer G, Wachter H, Mayer B (1995) Biosynthesis of nitric oxide: dependence on pteridine metabolism. Rev Physiol Biochem Pharmacol 127:97–135

Werner-Felmayer G, Golderer G, Werner ER (2002) Tetrahydrobiopterin biosynthesis, utilization and pharmacological effects. Curr Drug Metab 3:159–173

Wicher D, Messutat S, Lavialle C, Lapied B (2004) A new regulation of non-capacitative calcium entry in insect pacemaker neurosecretory neurons. Involvement of arachidonic cid, NO-guanylyl cyclase/cGMP, and AMP. J Biol Chem 279:50410–50419

Wideman RF, Erf GF, Chapman ME (2005) Nω-nitro L-arginine methyl ester (L-NAME) amplifies the pulmonary hypertensive response to microparticle injections in broilers. Poult Sci 84:1077–1091

Wiesinger H (2001) Arginine metabolism and the synthesis of nitric oxide in the nervous system. Prog Neurobiol 64:365–391

Wilken M, Huchzermeyer B (1999) Suppression of mycelia formation by NO produced endogenously in Candida tropicalis. Eur J Cell Biol 78:209–213

Wink DA, Osawa Y, Dabyshire JF, Jones CR, Eshenaur SC, Nims RW (1993) Inhibition of cytochrome P450 by nitric oxide and a nitric oxide-releasing agent. Arch Biochem Biophys 300:115–123

Xiao Q, Xu HY, Wang SR, Lazar G (2000) Developmental changes of NADPH-diaphorase positive structures in the isthmic nuclei of the chick. Anat Embryol (Berl) 201:509–519

Xu H, Huang K, Gao Q, Zhu Y, Han X (2001a) Prevention of axial elongation in myopia by the trace element zinc. Biol Trace Elem Res 79:39–47

Xu H, Huang K, Gao Q, Gao Z, Han X (2001b) A study on the prevention and treatment of myopia with nacre on chicks. Pharmacol Res 44:1–6

Yamakawa Y, Shiraishi H, Yamakawa M (1997) Characterization of nitric oxide synthase in the cerebellum of the chicken. Comp Biochem Physiol B Biochem Mol Biol 118:457–461

Yi GB, McClendon D, Desaiah D, Goddard J, Listr A, Moffitt J, Meer RK, deShazo R, Lee KS, Rockhold RW (2003) Fire ant venom alkaloid, isosolenopsin A, a potent and selective inhibitor of neuronal nitric oxide synthase. Int J Toxicol 22:81–86

Yonamine CM, Troncone LR, Camillo MA (2004) Blockade of neuronal nitric oxide synthase abolishes the toxic effects of Tx2–5, a lethal Phoneutria nigriventer spider toxin. Toxicon 44:169–172

Yu D, Eldred WD (2003) GABAA and GABAC receptor antagonists increase retinal cyclic GMP levels through nitric oxide synthase. Vis Neurosci 20:627–637

Yu D, Eldred WD (2005) Nitric oxide stimulates γ-aminobutyric acid release and inhibits glycine release in retina. J Comp Neurol 483:278–291

Yuda M, Hirai M, Miura K, Matsumura H, Ando K, Chinzei Y (1996) cDNA cloning, expression and characterization of nitric-oxide synthase from the Salivary glands of the blood-sucking insect Rhodnius prolixus. Eur J Biochem 242:807–812

Zaccone G, Ainis L, Mauceri A, Lo Cascio P, Lo Giudice F, Fasulo S (2003) NANC nerves in the respiratory air sac and branchial vasculature of the Indian catfish, Heteropneustes fossilis. Acta Histochem 105:151–163

Zagvazdin Y, Fitzgerald ME, Reiner A (2000) Role of muscarinic cholinergic transmission in Edinger-Westphal nucleus-induced choroidal vasodilation in pigeon. Exp Eye Res 70:315–327

Zagvazdin YS, Fitzgerald ME, Sancesario G, Reiner A (1996) Neural nitric oxide mediates Edinger-Westphal nucleus evoked increase in choroidal blood flow in the pigeon. Invest Ophthalmol Vis Sci 37:666–672

Zipfel B, Schmid HA, Meissl H (1999) Photoendocrine signal transduction in pineal photoreceptors of the trout. Role of cGMP and nitric oxide. Adv Exp Med Biol 460:79–82

Rev Physiol Biochem Pharmacol (2006)
DOI 10.1007/112_0602

S. Huang · J. Wikswo

Dimensions of systems biology

Published online: 12 September 2006

Abstract Systems biology, possibly the latest sub-discipline of biology, has arisen as a result of the shockwave of genomic and proteomic data that has appeared in the past few years. However, despite ubiquitous initiatives that carry this label, there is no precise definition of systems biology other than the implication of a new, all-encompassing, multidisciplinary endeavor. Here we propose that systems biology is more than the integration of biology with methods of the physical and computational sciences, and also more than the expansion of the single-pathway approach to embracing genome-scale networks. It is the discipline that specifically addresses the fundamental properties of the complexity that living systems represent. To facilitate the discussion, we dissect and project the multifaceted systems complexity of living organisms into five dimensions: (1) molecular complexity; (2) structural complexity; (3) temporal complexity; (4) abstraction and emergence; and (5) algorithmic complexity. This "five-dimensional space" may provide a framework for comparing, classifying, and complementing the vast diversity of existing systems biology programs and their goals, and will also give a glimpse of the magnitude of the scientific problems associated with unraveling the ultimate mysteries of life.

Introduction

The spate of "systems biology" initiatives that in the past few years has produced new institutions, departments, and research or educational programs suggests that "systems biology" is

S. Huang (✉)
Harvard Medical School, Department of Surgery and Vascular Biology Program,
Karp 11-212, Children's Hospital, 300 Longwood Avenue, Boston, 02115 MA, USA
e-mail: Sui.Huang@tch.harvard.edu · Tel.: +1-617-919-2393

J. Wikswo
Vanderbilt University, Vanderbilt Institute for Integrative Biosystems Research and Education,
Departments of Biomedical Engineering, Molecular Physiology & Biophysics and Physics & Astronomy,
VU Station B 351807, Nashville, 37235 TN, USA

a reality, here to stay, and not merely a buzzword. But what is systems biology? While there is currently no consensus on a definition, life scientists have surprisingly quickly adopted this new term into their vocabulary because, on the surface, it has conveniently allowed them to communicate the dawn of a new era of interdisciplinary, all-encompassing biology in academia and the biotechnology industry. The deeper reason for the rapid and unanimous acceptance of the term despite the lack of a tangible definition is perhaps that it reflects an increasing awareness of the limitations of "reductionism" in molecular biology and of the necessity to overcome them.

The vast diversity of justifications for systems biology initiatives mirrors the similarly wide spectrum of what individual life scientists regard as "reductionism." Some biologists are disturbed by the overt emphasis on genetic determinism (Strohman 1997; Morange 2001; Rose 2003). Others are more concerned about the neglect of quantitative analysis (Endy and Brent 2001), formalization, and abstraction, or the disregard of complexity and context (Goodwin 2001; Lewontin 2001). Because of this plurality of firmly anchored perceptions of molecular reductionism of the past decade and of opinions about what to do next, an attempt at an epistemologically precise definition of systems biology would be doomed to fail.

Certainly, systems biology is more than the return to traditional organismal physiology, as a few cynics have voiced. Molecular biology's revelation of the inner secrets of life has opened a new vista on living systems that distinguishes today's system-wide approach from that of the "innocent" days of classical physiology. Systems biology also is more than the logistic integration of bioinformatics databases and of computational modeling with experimentation. If the whole is more than the sum of its parts, as Aristotle taught us (*Metaphysica*), then the rapidly increasing knowledge of the (molecular) parts, stimulated by the "omics revolution" (Evans 2000; Ge et al. 2003) begs for an explanation of how the whole (organism) arises from its parts—other than by linearly adding them up. Therefore, one overarching goal of systems biology could be: the analysis of entirety rather than the entireness of analysis. This could be an operational definition were it not too philosophical to be useful. More concretely then, if systems biology is the migration away from the analysis of the individual molecular parts in isolation toward embracing the bigger picture of the components functioning as integrated parts of a whole, then its very nature will depend on which facet of reductionism is being evaded, as well as which aspect of complexity of the organism as a "whole" is being addressed. For this reason it is useful to dissect the fundamental nature of the complexity of living systems into naturally identifiable domains. Here we propose five principal aspects of living systems that not only represent fundamental, immanent properties of complex systems, but also roughly incorporate the operational goals of typical systems biology approaches. The proposed scheme provides a guiding reference system along five axes to help organize the diversity of existing initiatives. Depending on the relative emphasis, any research effort labeled "systems biology" can be projected onto the "hyperspace" spanned by each of the five dimensions (see Table 1): (1) molecular complexity; (2) structural complexity; (3) temporal complexity; (4) abstraction and emergence; and (5) algorithmic complexity. These five component aspects are not mutually exclusive but might overlap, so that as formal dimensions they are not perfectly orthogonal.

Table 1 The five dimensions that span the complexity of living systems and the challenges of systems biology

Dimension	Research objectives	Approach, tools or traditional disciplines	Challenges
1. Molecular complexity	Integrating molecules and pathways to genome-wide networks ("horizontal integration")	Molecular biology, genomics, proteomics, and other "omics," bioinformatics, *applied* computer science	Database integration, annotation; high-throughput analysis at higher functional levels; real-time, multiplex measurements of expression profiles
2. Structural complexity	Transcending many size-scales (nanometer to meter): Organelles–cells–tissues–organs–organism ("vertical integration")	Microscopy, nanotechnology, microfabrication, biophysics, biomedical engineering, mechanical engineering, anatomy	Visualization of dynamic protein complexes in cells; cellular compartmentalization; external sensing and actuation of molecular states; bridging the gap between macromolecules and tissues—spanning nanometer to centimeter scales
3. Temporal complexity	Transcending many time scales (nanoseconds to gigaseconds = decades)	System dynamics, nonlinear dynamics, signal processing, biomedical engineering, control theory, time-series analysis, physiology	Noise measurement at various levels; non-linear time series analysis; multi-analyte measurements in non-equilibrium systems—spanning nanosecond to gigasecond time scales
4. Abstraction and emergence	Modeling system-level "emergent" features	System dynamics, pattern-formation, cybernetics, complex system sciences, agent-based models, mathematics	Levels of abstraction, multi-scale models and scaling; use of larger-scale effective variables to describe smaller-scale phenomena; "statistical mechanics of biology"
5. Algorithmic complexity	Understanding information coding and computation by the biological medium and developing models that simulate biological systems	*Theoretical* computer science, information theory, electrical engineering	Identifying a core system that is computationally irreducible; new syntax for simulation; the integration of hybrid analog/digital models

Dimension 1: molecular complexity

After decades of genetic analysis, the most obvious aspect of complexity in living organisms appears to be the large but finite number of genes and proteins in the genome of an organism, which stimulates the ambition to "know them all." Thus, one of the earliest departures from molecular reductionism was driven by the notion that understanding life processes will first require knowledge of all the parts of an organism, not just individual molecules in isolation. This unquestioned idea was fostered by the arrival of genomics and proteomics, which have quickly triggered other "omics" sciences. Technologies for massively parallel and high-throughput analysis, such as automated DNA sequencing and nucleotide and protein microarrays, as well as advances in bioinformatics and database integration, brought the ambitious goal of a systematic—if possible, exhaustive—identification, categorization, and characterization of all genes and proteins of a genome within reach. This was to be followed by the study of the functional aspects of individual genes and their encoded proteins, such as the interaction partners, biochemical activities, and biological role.

However, such "functional genomics" approaches, as they were initially called to distinguish them from traditional sequence-centered genomics, nonetheless often appeared as "brute-force reductionism," in which technology was used merely to accelerate the comprehensive characterization of pathways. In such massively parallel reductionism, entireness of analysis still trumps understanding of entirety. The significance of molecular complexity arises not only from the variety of individual molecules themselves, but also from how they function in consort. It is becoming obvious, as genomes of model organisms are sequenced, that functions of living systems are determined not only by individual genes and their encoded proteins, but also by how evolution led to an ever-increasing complexity of protein–protein and other intra- and intercellular interactions. The "organizational complexity" of these interactions (Strohman 2000), rather than punctual differences in the genome sequence, distinguishes higher organisms from lower ones. This is most prosaically manifested in the surprising (perhaps disappointing) finding that the number of genes in humans (around 25,000 in the latest count) is in the same order of magnitude as in disproportionally more "primitive" organisms, such as the plant *Arabidopsis thaliana* or the worm *Caenorhabditis elegans* (Southan 2004). Even if we consider differential splicing and posttranscriptional and translational modifications, as has been argued in an effort to restore the molecular superiority of humans (Graveley 2001), it is not at all clear whether we can directly map genomic complexity into phenotype complexity. Instead, it is obvious that the regulatory interactions between genes, proteins, and metabolites contribute, through combinatorics and constraints, to the enormous richness of this dimension of molecular complexity, and hence, ultimately can provide an answer for why the whole is more than the sum of the parts. In fact, the key operation is multiplication rather than addition—the whole is the product of high-order combinatorial multiplication, not a simple linear summation.

Consequently, one of the first explicit goals of "post-genomic biology" is to establish, piece by piece, the genome-wide "network map" of all the specific, regulatory interactions between the molecules—the wiring diagram, so to speak—of an organism (Davidson et al. 2002). Such endeavors brought the notion of "complex networks" (Marcotte 2001; Strogatz 2001; Barabasi and Oltvai 2004) into molecular biology. The biology of networks was a first step toward conceptualization beyond the ad hoc models in modern biology in which molecular pathways represent chains of causation and hence provided an explanation of macroscopic processes. Some experimentalists now even equate systems biology with "network biology." Unraveling the molecular wiring diagram can be achieved in essentially two ways: the painstaking demonstration of individual interactions through experiments, or the use of a formalism to reverse engineer the network structure of a system based on system-wide measurement of the coordinated dynamic of molecular activities that is governed by the network interaction (D'haeseleer et al. 2000; Gardner et al. 2003). The study of the architecture of genome-wide interaction networks has stimulated an avalanche of theoretical work—not only in biology but also in physics (discussed in Sect. 5) (Barabasi and Albert 1999). However, the identification of topological features of a static network's topology features is only one of the first steps on the path to understanding organismal complexity. The network architecture imposes a constraint on how individual elements at the network nodes, the genes and proteins, can behave. Hence the next step is necessarily the study of the dynamics of the collective behavior of the molecules in the network. This will be discussed in Sects. 4 and 5 (Dimensions 3 and 4).

The need for the computational capacity to analyze and manage the vast amount of sequence, protein structure, functional, and expression profile data has brought bioinformatics to the center stage of genomics, necessitating the integration of wet lab work with computational science. However, although this cross-disciplinary integration of methodology is most apparent in this dimension of molecular complexity, computation in biology is more than a mere tool to handle large amounts of data. Computing is also a tool for modeling and simulation, as is discussed in Sects. 3, 4, and 5 (on Dimensions 2, 3, and 4). But beyond that, in a more profound sense, computing is an elementary process of complex systems itself, including the living organism, and hence it is an object of, rather than a tool for, analysis, as we will encounter in Dimension 5 of systems biology.

But in this first dimension, the integration of computation serves the default goal of the "omics" sciences, which is centered around identification, description, characterization, and classification of molecules, and hence appears to some critical observers as the molecular equivalent of the "stamp collecting" of old-time botany and zoology (to paraphrase Rutherford and Watson) (Blackett 1963; Wilson 1995). Even if the study of molecular networks now embraces quantitative analysis following targeted perturbations (Ideker et al. 2001), the systematic quest for the missing parts of a finite genomic jigsaw puzzle is governed by the same kind of mind-set that guided the great explorers of the past to the discovery of new islands or rivers on our finite planet. This mentality is reflected in the original perception of systems biology as a "discovery" science as opposed to the traditional "hypothesis-driven" sciences (Aebersold et al. 2000; Brent 2000; Ideker et al. 2001). The underlying motivation for the brute-force characterization of all the molecular parts and connections is the subconscious but widely held belief that knowing all the component parts of a system and their wiring diagram is equivalent to understanding that system. The problem is that this constructionist approach, thought to be a reversion of reductionism, preserves the fallacies of reductionist thinking (Anderson 1972).

Modern-day molecular stamp collecting, however, is warranted and necessary. The set of stamps is large, but finite, and it is not unrealistic to hope to collect them all within the next few decades. The comprehensive characterization of all quantitative and qualitative details is important because of one essential aspect of the complexity of living systems: the immense heterogeneity of the individual components and the astronomic combinatorial possibilities in which these parts (genes, peptides, cell types, organisms) can interact to generate function (Elsasser 1998). This unfathomably large "space" that arises from the combinatorics makes evolution a (quasi) non-ergodic process, so that not all features of organisms may be the result of functional optimization by adaptation (Dawkins 1996; Alon 2003), as an engineer would like to see it (Mangan and Alon 2003). Instead, organisms are replete with features that may represent frozen historical accidents (Gell-Mann 1995; Gould 2002) and local optima fixed by evolution (Kauffman 1993). Moreover, natural selection had to face the constraints imposed by laws of physics, rules of architectural design and the principle of evolvability (Gould and Lewontin 1979; Webster and Goodwin 1984; Autumn et al. 2002). Conversely, it can sometimes get a free ride from self-organizing processes that can generate more complex ordered patterns without external instruction (as discussed in Sect. 5) (Kauffman 1993), which in turn can be selected for. All this reduces the role of the optimizing (adaptive) force of natural selection and produces idiosyncrasies that are often difficult to capture by universal principles. The end result is a living organism which runs on software (the genome) that consists of multiple layers of "clever hacks," invented ad hoc but kept forever because they work, rather than on a software that has been systematically designed de novo by disciplined programmers and optimized for algorithmic elegance or efficiency. The "organized heterogeneity" (Elsasser 1998) of an immense variety of subparts is a fundamental characteristic of living organisms that is absent in physical systems, and the dichotomy between idiosyncrasy and universality poses a challenge to formalization of multicomponent living systems in life sciences, in contrast to the description of homogeneous multicomponent nonliving matter in statistical physics. Thus, a systematic, exhaustive analysis of the individual parts, their interaction modes and quantitative parameters, as summarized in this molecular dimension of biological complexity, remains an important task in future biology. Indeed, although to early advocates of integrative biology the analysis of gene sequences was the poster-child of reductionist thinking, modern bioinformatics of (whole) genome sequence comparison between species has become one of the most important and exciting tools for the study of evolution (Medina 2005). More recently, such "brute-force," large-scale comparisons have been extended to gene expression profiles, thus adding a functional aspect (Stuart et al. 2003). Holistic minds in biology will contend, however, that the availability of an entire list of parts and their functional specifications is but the first stage in the journey toward understanding a given living organism as an entity, a system. Therefore, other dimensions of complexity need to be considered.

Dimension 2: structural complexity

While the genomic perspective of complex systems emphasizes the information-encoding and processing aspect of interacting biomolecules, i.e., the basic software, the complexity of a living organism is also manifested in a characteristic fashion in the structure of its hardware. To understand how the information in molecular networks materializes into the mechano-electro-chemical machinery of organisms, we need at least to understand the actual underlying "device physics." This aspect of a living system has been transparent to the genomic sciences, notably bioinformatics, but was the domain of traditional biophysics. So what new perspective in the study of the physics of biological devices then warrants the label "systems biology"?

A characteristic aspect of complexity of living systems is that they exhibit feature richness at multiple size scales across a wide range; i.e., they have a deep "scale space" (Bar-Yam 2004). Genes determine proteins (at the nanometer scale), which organize the chemistry so as to assemble organelles (micrometer) and cells (~10 μm). A group of cells assembles to tissues with characteristic patterns (millimeter). Tissues form organs (centimeter), and organs cooperate within the systemic physiology of the organisms (meter). In other words, when looking at a living organism through an imaginary universal magnifying device, one could continuously zoom in and out without encountering the abrupt jumps in feature richness that are often found in inanimate materials of man-made systems. (Note that a feature boundary such as the nuclear or mitochondrial membrane represents a spatial boundary but not necessarily a jump in feature richness.) The smooth variation in feature size in biological systems, spanning nanometers to meters, is a fundamental property that distinguishes living systems from man-made machines.

It is important to note that this hierarchical, multiscale complexity in the structure of higher organisms is a more encompassing principle than fractals, which in addition to feature richness at all size scales also exhibit self-similarity between the scales (Mandelbrot 1982; Bassingthwaighte et al. 1994; Cross 1997). Fractals are found in some subsystems within a complex organism, such as in bronchial or blood vessel branching systems, but these physical fractals extend over a finite range of size scales. Moreover, the defining property of fractals, the scale-invariance (self-similarity at different scales) (Mandelbrot 1982), is distinct from the multiscale feature richness of complex organisms discussed here. The latter is characterized by scale-specific features that vary between different levels of organization (molecules, cells, tissues) and are not necessarily repeated at other size scales. If the complexity discussed in Dimension 1 epitomizes the proposal of organized heterogeneity as a characteristic feature of living systems (Elsasser 1998), then the structural complexity discussed here may be summarized as the principle of (heterogeneous) hierarchy (Pattee 1973), which provides a second characteristic organizing principle.

Importantly, in this nested hierarchy each level exhibits system properties not obvious from the properties of its component parts (e.g., proteins) and harbors its own set of rules that determine new interaction modalities (e.g., protein–protein interaction, cell–cell communication) not evident at the smaller scale. Or, as Phil Anderson put it, psychology is not applied biology, biology is not applied chemistry, and chemistry is not applied physics (Anderson 1972). This apparent "irreducibility" of higher-level features is sometimes, without rigorous definition, referred to as an "emergent property." Hence, multiscale structural complexity of higher organisms is tightly linked to emergence and abstraction, which represent our Dimension 4 of complexity.

Given the scale-transcending, hierarchical property, the study of structural complexity entails taking a multiscale approach, rather than studying each level separately—as was done in traditional biophysics. The depth of scale space required to fully describe physiological phenomena represents a key manifestation of structural complexity. In the heart, for example, the stability of the cardiac rhythm is determined by how wave fronts propagate over many centimeters, but this in turn is determined by features as small as ion channels, gap junctions, and intercalated disks in the nanometer range (Wilders and Jongsma 1993). In an individual cell, the micron-scale spatial localization of a signaling event (Simon and Llinas 1985) would suggest the need for submicron spatial descriptions of cellular activity, a point amply illustrated by high-resolution images of intracellular heterogeneity (Marsh et al. 2001). Structural complexity is thus already evident at this lowest molecular level: So-called molecular crowding (Hall and Minton 2003) of the cytoplasm leads to different scales of diffusion rates, precluding the use of classical kinetic formalism derived for ideal, well-stirred, single-scale homogeneous solutions. This necessitates the development of more explicit multi-algorithm modeling (Takahashi et al. 2005). From this perspective, a description of the state of a macroscopic organism in terms of the state of each independent submicron-to-micron feature would require the impossible specification of what could be a "mole of parameters"—the true measure of structural complexity.

If large-scale genomics (Dimension 1) was the horizontal integration of traditional molecular biology approaches, then transcending the various hierarchical levels of structures within one research project can be viewed as a vertical integration. What are the tools for this approach? In the molecular genetics of Dimension 1, the typical experiment consists of a genetic or pharmacological perturbation of a given system, followed by the (genome-wide) analysis of its response using established molecular biology and biochemistry. When studying structural complexity and associated "emergent features" (discussed in Dimension 4), perturbation translates into actuation and control, and the equivalent of molecular analysis is sensing, not only of changes in activity of molecular species as carriers of information, such as protein phosphorylation state, but of physical variables, such as spatial-temporal distributions of pH, temperature, photons, currents, and voltage, as well as shapes, motion, mechanical forces, material properties, etc., at the various levels of organization. In the horizontal dimension of molecular complexity, broad-scale and high-throughput tools have long been available, but the toolbox for climbing up the ladder of structural integration toward larger scales is almost empty. A systematic effort to develop standard devices for actuation, control, and sensing that would allow the manipulation of more complex experimental systems (cells, tissues, organoids) and the measurement of higher-order, system properties at each scale will be paramount for studying this aspect of complexity. As we will see in the next section, the ability to control biological processes at all spatial scales will also be critical for understanding the dynamics that result from the third dimension—temporal complexity.

Dimension 3: temporal complexity

The multiscale property of complexity applies not only to space, as discussed above, but also to time, the dimension of dynamic behavior. Chemical and physical processes exhibit characteristic temporal structures. While electron transfer reactions can occur in femtoseconds, protein conformational changes associated with cellular signaling are slower, taking place in nanoseconds. The nature of rapid ion channel gating events (microseconds), such as those that determine the depolarization of the heart (milliseconds), in turn determine the stability of the cardiac cycle (1 s), and can affect directly the longevity of the organism (gigaseconds), for a total span in time of a factor of 10^{18}. In comparison, the most complex astrophysical calculations to date, which describe the gravitational accretion that led to large-scale structure in the universe, may span only a factor of 1,300 in time and involve only one force (gravity) and six degrees of freedom for each node (position and velocity) (Springel et al. 2005).

The most elementary complex temporal "pattern" (in the widest sense of the word) is fluctuations of variables due to stochastic processes, such as the local intracellular level of a protein that fluctuates in time because of the effect of finite numbers of molecules in the cells (e.g., two gene copies, 1,000 s of transcription factor copies) (Kaern et al. 2005). The study of "gene expression noise," although its cell-biological manifestations in higher organisms have long been recognized (Enver et al. 1998; Hume 2000), has gained momentum in the era of systems biology, perhaps because the probabilistic nature appealed in a particular way to those who sought to escape genetic determinism.

But the more prosaic manifestations of temporal complexity are patterns in a narrower sense of the word, namely those that the human brain intuitively perceives as "regular," for example, oscillations. Such temporal patterns of behavior "emerge" from nonlinear regulatory interactions (stimulus-response relationships), including thresholding, saturation and "biphasic" (non-monotonical) effects (Tyson et al. 2003), which connect genes, proteins, and metabolites to establish circuits and networks. A central goal in studying the generation of the temporal patterns in the behavior of a system is to understand the formal relationship between the network architecture on the one hand—which can implement a breadth of feedback, modulation, and control schemes well known to system engineers—and the observed dynamics on the other hand. Major research agendas include either reverse engineering to infer the network structure based on observed dynamics (as mentioned in Dimension 1) or the modeling and simulation of a given network structure to learn about the dynamic behavior the network governs.

Temporal oscillations, the most basic temporal pattern, in gene expression are best known in the cell division cycle, but also occur in a variety of other cellular processes, such as in the change of redox state or in the cellular response to DNA damage (Klevecz et al. 2004; Lahav et al. 2004). Periodic dynamics also are observed for higher-level physiological variables, including transmembrane potentials, cytosolic calcium, secretion of humoral mediators, neural activities, and muscular contractions (Glass 2001). Oscillations can also be local manifestations of large-scale spatiotemporal traveling waves, such as waves of excitation in the brain or heart (Gray et al. 1995) or of cellular behavior as in *Dictyostelium* (Palsson and Cox 1996), animal swarms, flocks, and schools (Okubo 1986) or even nationwide infectious outbreaks (Cummings et al. 2004). Such spatiotemporal patterns at various scales of magnitude are elementary but particularly impressive examples of "emergence" as discussed in Dimension 4.

Temporal patterns that arise from deterministic interactions and constraints of a network, however, need not be periodic but can also appear irregular, as in the case of deterministic chaos, which must be distinguished from the stochastic fluctuations discussed above (Mackey and Glass 1977; Gleick 1988). The discovery and study of chaotic behavior—as well as its spatial equivalent, the fractal structures—in life sciences (Goldberger et al. 2002) was perhaps one of the earliest and starkest demonstrations of the limits of "linear thinking" in mainstream molecular genetics. The "dynamic explanation" of disease defied the traditional reduction to genetic pathways. Chaotic behavior, e.g., in heartbeat frequency, has been associated with the healthy state, whereas loss of this type of temporal complexity is observed in disease states (Goldberger et al. 2002). Today, one might readily conclude that chaos and fractals, which can occur at a variety of size and time scales, are properties of a rather small set of subsystems within the complex system of the living organism. However, this conclusion may be based upon inadequate data: while the regulatory dynamics of the cardiac rhythm has been well studied because the heart rate is easily measured, the fractal nature of heart rhythm has only recently been appreciated (Ivanov et al. 1999; Goldberger et al. 2002), in part because of technical challenges in the numerical analysis of long-term time-series data. We should be prepared to uncover new dynamics as we begin to explore single-cell signaling, as with the recently discovered pulsatile expression of p53 following DNA damage to single cells (Lahav et al. 2004). The observation of multifractality in heartbeat dynamics raises the intriguing possibility that the associated nonlinear control mechanisms involve coupled cascades of feedback loops in a system operating far from equilibrium (Ivanov et al. 1999; Goldberger et al. 2002). Were this to prove true for larger classes of biological regulatory systems, one might encounter unexpected difficulties in creating models based on reverse engineering of cellular signaling and regulation.

Temporal complexity is not only manifested in the measured temporal pattern of a variable (e.g., spikes vs sinusoid vs chaotic), but also in the stimulus-response characteristics: in the simplest case of linear sensors, a temporal shape of a stimulus (e.g., a "rectangular stimulus") may be mapped directly into the response shape. But the living system may also just "feel" the change rather than the actual intensity of a signal, i.e., sense its first time derivative (e.g., one spike in response to a step-shape change). Such differential sensors are important for adaptation (habituation, wherein we don't feel our clothes) and alerting the organism to sudden changes in the environment. There is also the need for a system to completely suppress the response to a change in environment (perfect adaptation, i.e., not even a spike in response to a step-shape change), a manifestation of robustness (Alon et al. 1999; Yi et al. 2000) (see Sect. 5). Another important facet of temporal response patterns is whether the response returns at all when the stimulus disappears (reversibility vs irreversibility). A characteristic of nonlinearity of the underlying mechanism is hysteresis, where the response strength to a given stimulus strength may depend on the history (previous state) of a system (Laurent and Kellershohn 1999; Ninfa and Mayo 2004). Hysteresis can be viewed as a primitive memory effect and is related to multistability and nongenetic persistence discussed in Sect. 5. Other aspects of temporal response characteristics include time delay, integration, and more complex signal transforms.

An important issue to note when analyzing temporal patterns is the interference with the vertical integration in structural complexity (Dimension 2): since tissues of multicellular organisms are composed of cells as functional units, tissue-level variables represent ensemble averages. The traditional monitoring of the change of expression of a gene in a tissue (e.g., as measured by RT-PCR, microarrays, or immunoblots of tissue lysates) produces an aggregate variable, averaged over the expression changes in all the individual cells in that tissue (Huang 2005). Unless the individual cells are synchronized with respect to the measured variables, such population measurements incur loss of information. As more and more (dynamic) measurements can now be made at the level of single cells, oscillations or other temporal patterns within individual cells that are masked by multicell ensemble-averaging will become evident (Xiong and Ferrell 2003; Levsky and Singer 2003; Lahav et al. 2004; Sachs et al. 2005). Tissue-level oscillations can be observed in the presence of either experimental or natural synchronization of the behavior of the individual cells (Whitfield et al. 2002; Klevecz et al. 2004; Yamamoto et al. 2004). Thus, the heterogeneity of a population of genetically identical cells, either due to random fluctuations of gene expression (Kaern et al. 2005) or persistent epigenetic individuality (Spudich and Koshland 1976; Rubin 1990; Ferrell and Machleder 1998; Balaban et al. 2004) must be considered. Complex organisms either exploit stochastic cell population heterogeneity to achieve a smooth, macroscopically continuous tissue behavior (e.g., in the graduated control of skeletal muscle contraction strength by varying numbers of activated fibers) or suppress it by coordinating cell behavior through structural or chemical cell–cell coupling (e.g., in generating the macroscopic pumping function in the heart). Such cellular ensemble behavior creates an "emergent temporal pattern" at a larger time and space scale, represented by another set of variables at a hierarchically higher level of structural organization. One challenge can be to identify the messengers that mediate the regional synchronization as well its function, as in the case of calcium oscillations in a single pancreatic islet (Rocheleau et al. 2004).

The notion of multivariate dynamic behavior in individual cells within a cell population adds another layer of complexity to deterministic genomics and poses new challenges for the acquisition and analysis of data. First and foremost, the study of temporal patterns will require the capacity for precise, real-time measurements of variables at differing time and size scales. In particular, systems will have to be studied when they are far from equilibrium: e.g., analysis of the dynamics of the metabolic response of cells to chemical perturbations (Eklund et al. 2004) will extend metabolic modeling that heretofore has concentrated on the steady-state analyses of metabolic fluxes. Passive acquisition of data alone is insufficient. It is well recognized that it is difficult to obtain a detailed understanding of a closed-loop, feedback system without opening the feedback loop—hence the intuitive emphasis in traditional biology on genetic perturbations, such as gene knockouts and RNA interference. However, it is not sufficient to measure dynamic variables following perturbations in such systems; it is also necessary to assert external control of the system to explore gain and stability of the feedback mechanisms (Wikswo et al. 2006). It is not yet widely recognized that the advances provided by the availability of more and more sensing capabilities for analyzing a single cell will be limited by the need for corresponding cellular actuators. At present, natural proteins that act as molecular machines by changing conformation or binding affinity in response to a stimulus can be thought of as actuators—albeit of limited controllability. But there is a growing need for externally fine-tunable subcellular actuators that can accomplish similar tasks. Thus, as is the case for the study of structural complexity, this will entail the development of micro- and nano-devices (Freitas 2002; Whitesides 2003) that act as intracellular switches, sensors (Lu and Rosenzweig 2000), or valves (Nguyen et al. 2005; Kocer et al. 2005) in combination with the latest achievements in microscopy and digital imaging techniques.

To integrate the analysis of the first three dimensions of systems biology covering the molecular, structural, and temporal complexity in living systems, our capabilities in gene- and protein-specific measurements (Dimension 1) need to be advanced and combined with measurements of the other two dimensions. Concretely, one of the ultimate goals would be the development of tools for high-resolution, real-time (Dimension 3), multiplex determination of individual molecular species (Dimension 1) within the context of the physical structure (Dimension 2), at various levels: from single cells to tissues to organs. To date, multiplexing measurements (e.g., of thousand-dimensional gene expression vectors by DNA microarrays) requires the destruction ("lysis") of biological structure, while real-time measurement of genes and protein states in the intact, ordered structure are low-dimensional. But the first, small steps to overcome this challenge have recently been taken, such as expression profiling in single cells (Levsky et al. 2002; Le et al. 2005).

Dimension 4: abstraction and emergence

Many phenomena associated with complex systems are better understood in terms of general, abstract principles without making reference to particular material constituents. To illustrate this, let us use the example of circadian rhythm. Its oscillatory nature per se is what we think is of interest from a systems perspective, and hence needs to be studied as such, in a generic and abstract sense (Barkai and Leibler 2000). In contrast, traditional molecular biology by default focuses instead on identifying the tangible, material substrate that underlies the observed phenomena, in this case, the particular genes "that are involved" in circadian rhythm (Lowrey and Takahashi 2004). The identification of genes shown to be critical for the periodicity may provide a mechanistic, qualitative ad hoc explanation for the existence of the rhythm. But such findings, while necessary, are not sufficient to understand how the more abstract property of oscillation arises from the organization and interplay of the individual components. More generally speaking, molecular biologists are mostly satisfied with a concrete and specific, so-called proximal, explanation ("The car moves because the wheels are turning") (Tinbergen 1952; Dretske 2000). This epistemological habit is manifest in the ubiquitous hunt for the molecular pathway that causes a particular behavior or disease. Were living organisms Rube Goldberg machines (Wolfe and Goldberg 2000), no further explanation would be needed. In contrast, physicists and mathematicians, and to a lesser extent, engineers, prefer to study such phenomena from a general perspective not closely tied to the underlying particulars ("The car moves because of the thermodynamic laws that the combustion engine has to obey"). The challenge for systems biology will be to unite these epistemologically disparate efforts and the associated communities.

In the by now traditional complex systems sciences (Waldrop 1992; Gell-Mann 1995; Goodwin 2001), the more abstract "system features," as exemplified above by the oscillatory behavior, are at the center of interest and have come to be known as "collective behavior" or, more poetically, "emergent properties," as encountered earlier. The latter term is not mathematically rigorously defined (Corning 2002) and some scientists question its necessity. Nevertheless, emergent properties have become somewhat a defining characteristic of a complex system (Bar-Yam 1997). The term is intuitively appealing to many and is operationally useful for communicating the idea of system-level features that cannot be understood by studying the dissected parts in separation. Here, given its widespread and loose usage, we propose that the term "emergent" refer to any abstract property of a system that is not obviously manifested as a property of an individual material subcomponent of the system. Thus, oscillations, switch-like behaviors, and striped or spiral wave patterns in an otherwise continuous system are examples of simple emergent features (Turing 1952; Murray 1993; Meinhardt 1996).

It is important to note that an emergent behavior in this definition is not equivalent to any notion of "irreducibility" in a stricter sense, be it the material, epistemological, or computational irreducibility (discussed in Sect. 10), nor is there anything romantic or magical about emergence, as is perhaps implied by the notion that the whole is more than the sum of its parts. In contrast, an emergent system property, in the loose definition used here, can in fact often be "reduced" to its underlying parts, i.e., understood as a consequence of the interplay of the parts. But to do so entails simplification, abstraction, conceptualization, and formalization. As philosophers of science and pedagogy put it, there is no understanding without abstraction, i.e., leaving out irrelevant details (Picht 1969).

The tool for abstraction in the widest sense is mathematical modeling, which comes in many forms and flavors. Mathematical modeling is in essence a shortcut that allows us to describe a system so as to predict its general or long-term behavior, given its structure and initial conditions, without having explicitly to "reenact" its behavior step by step from the knowledge of the specific parts and rules of their interaction. The best-studied emergent properties are features of dynamic behavior, and the tools used to describe such behavior in formal terms, typically differential equations, are well established through the discipline of system dynamics and control theory. Herein, the simplest emergent properties in living systems are dynamic features, such as temporal oscillations or stable or unstable steady-states, in particular, multistability. The last implies the existence of multiple alternative oscillatory or stable states, or "attractors" within one system (see below). On the other hand, as we have seen above, emergent properties can also be manifested as the more tangible geometric patterns in physical space. Generally speaking, we can say that emergence of both abstract behaviors and concrete structural patterns typically involves the counterintuitive transformation of gradual, quantitative differences into discrete, qualitative differences (Anderson 1972).

Emergent dynamic behaviors

In fact, the abstract dynamic behavior (temporal patterns) that emerges as a distinct quality can be represented as a geometric quality. To do so we need to introduce the concept of an abstract space, the so-called state space, which contains all possible states that a dynamic system can occupy, e.g., all gene expression patterns within a cell, if we model a gene network. The state space offers a more intuitive grasp of the behavioral repertoire of a system. The mathematical concept of the state space is central for the understanding of the potential behaviors of a multicomponent system, such as a molecular network. For instance, the presence of multiple stable states in a gene regulatory network would lead to the compartmentalization of a state space into "basins of attraction" in that all the unstable system states in the neighborhood of a stable "attractor" state would be attracted to it. Thus, attractor states illustrate how the very same gene network generates a variety of stable gene expression states representing the cell types in metazoan (Kauffman 1993; Huang 2005). The intuitive character of the state space representation is nicely illustrated in Waddington's metaphor of the "epigenetic landscape" (Waddington 1956; Reik and Dean 2002) in which watersheds, hills, and valleys compartmentalize a cell's state space and in doing so, represent a cell's fate decisions and development into stable cell types (the attractors, or valleys). It is important to realize that the most compact representation of the key features of a state space can, in some cases, be in the form of effective variables that have no exact physiological correlate, but provide a remarkable analytical mechanistic explanation of the phenomena, as with the "activation" and "inactivation" variables in a simple two-variable model of the nerve action potential (Fitzhugh 1961). Similarly, an explanation of trajectories in state space using the first few principal components can be more informative than the thousands of variables that are combined to produce these components (Huang et al. 2005).

Few molecular biologists have fully embraced the idea of a state space to conceptualize dynamic behavior. Instead, they mostly operate in the domain of network architectures (topologies) as evidenced by the preoccupation with pathways and network charts as explanatory schemes. Multistability, which is beyond such proximal explanations, establishes a system behavior with a memory that can epigenetically (without involving changes in the genome sequence) store and remember environmental inputs (Xiong and Ferrell 2003; Ozbudak et al. 2004), and offers a formal underpinning for the long-recognized phenomena of nongenetic inheritance, enduring modifications, and other epigenetically persistent states (Spudich and Koshland 1976; Rubin 1990; Balaban et al. 2004).

Pattern formation

In contrast to these rather abstract emergent properties, what has long attracted interest among organismal biologists, and more recently, complex system scientists, is the spontaneity or inevitability (given a certain system structure and initial condition) of how spatiotemporal patterns, such as stripes, spirals, and waves, as briefly mentioned in Sects. 3 and 4 (Dimensions 2 and 3), arise (Meinhardt 1996). At the core of spontaneous pattern formation are processes that resist the pull toward "indifferent randomness" or symmetry (thermodynamic equilibrium), and hence are also called symmetry-breaking events in a wider sense (Kirschner et al. 2000; Sohrmann and Peter 2003). Spontaneous pattern formation is also referred to as "self-organization" (Ball 2001)—a term equally poetic, overused, and poorly defined as "emergence." The attempt to understand the spontaneous generation of order as a process that occurs far from thermodynamic equilibrium led to the (still controversial) idea of "dissipative structures" (Nicolis and Prigogine 1989), i.e., quasi-stationary, ordered structures that take up (nonthermic) energy and produce entropy to maintain order. Patterns (e.g., tissue structures, coat patterns) at a higher level are created by the particular (typically nonlinear) behaviors and interactions of the components at the lower level (e.g., molecules, cells) in time and space, and have characteristic lengths that are usually orders of magnitude larger than those of their constituent components (Turing 1952; Murray 1993;

Meinhardt 1996); thus, pattern formation can be viewed as one of the natural processes that create the depth of scale space and hierarchies in Dimension 2 of complexity.

Other forms of emergence in molecular biology

The idea of emergence of abstract entities at higher levels of organization in living organisms is not limited to classical dynamic systems and pattern formation. The analysis of genome-wide gene regulatory or protein interaction networks, for instance, has benefited from "abstracting away" the particulars of molecular details (Dimension 1) to reveal another kind of emerging properties. This has opened the possibility of using graph theory for the analysis of network architectures, which has led to the discovery of particular "emergent" structures in the global architecture of the wiring diagram (Barabasi and Albert 1999; Barabasi and Oltvai 2004) or of local network motifs (Milo et al. 2002) in biomolecular networks. However, the functional and evolutionary significance of these graph theoretical properties remains to be studied carefully in the context of dynamics.

A central aspect of emergence in living systems is homeostasis, which is related to the more abstract property of the aforementioned robustness (to perturbations) (Yi et al. 2000; Stelling et al. 2004) that arises from the particular structure of the underlying wiring diagram between the component parts. The intricate interplay between robustness (stability), flexibility (adaptability), fragility (vulnerability), and complexity is at the core of living systems, behind the apparent temporal behavior (Goodwin et al. 1993; Carlson and Doyle 2002; Kitano et al. 2004), yet these properties are not much studied in the current "mainstream" complex systems approaches. [Robustness of a dynamic system should not be confounded with the use of "robustness" in a more obvious and concrete way in the context of graph properties and the analysis of network topology discussed in Sect. 2 (Dimension 1). There, robustness is not the ability of maintaining a system state in response to perturbations, but of a graph (network topology diagram) to maintain some global structural connectivity property in response to structural destruction of network nodes or links (Albert et al. 2000).]

Another fundamental and truly concrete emergent property in biological systems pertains to the question of how the soluble biochemistry of genes and proteins studied in Dimension 1 produces the tangible, macroscopic living systems (which exhibit the spatial structures of Dimension 2 and the temporal ones analyzed in terms of Dimension 3) that are subjected to the laws of macroscopic mechanics. How does solution chemistry create form and physicality of macroorganisms? Again, high-throughput characterization of all physical parts and their relationship alone will not suffice to answer this question. One approach is the formalization of physicality in the tensegrity (=tensional integrity) model for studying the emergence of macro-mechanical properties in living organisms (Ingber 1998; Ingber 2003). The concept of tensegrity explains how systems produce "macroscopic" mechanical stiffness and shape-stability through the particular spatial organization of "microscopic" tensile and compression resistant elements in a physical network, arranged so as to produce global force balance. Such a design principle has been suggested to be embodied in the interactions between amino acid residues within folded proteins, between the molecular filaments of the cytoskeleton within the cell, or between muscles, tendons, and bones in the musculoskeletal system (Ingber 1998; Ingber 2003). It allows a system to unite both mechanical stability and flexibility.

Spectrum of models of varying abstraction

It would go beyond the scope of this article to review the various types of modeling and the levels of abstraction that help formalize "emergence" in biological systems. But for this discussion of systems biology initiatives, it suffices to point to the broad spectrum of abstraction in the various biological modeling approaches, with extremes at the opposite ends that are complementary to each other (Huang 2004). At one end are the specific, detailed models that use stochastic equations that incorporate the behavior of single molecules, or with some minimal abstraction, deterministic equations that describe the kinetics and flows of molecular species (for modeling cellular processes) or cell types as entities (for modeling tissue processes). At the other end are models of such considerable abstraction as to be almost inconceivable to experimental biologists, such as cellular automata, graph theory, or even game theory.

Systems biologists with molecular biology backgrounds and engineers engaged in biology typically populate the "specific, detailed modeling" end of this spectrum and seek to design models as detailed and realistic as possible in order to "predict the dynamic behavior" of a given instance. Here the underlying conceptualization, such as formal reaction kinetics and laws of mass action, are not questioned, but even taken for granted, and all that is needed, so goes the idea, are precise experimental measurements of the quantitative parameters that can then be plugged into the models to validate them. A historical example of this detailed modeling based on precise quantitative data is the Hodgkin–Huxley model of the action potential. Such models based on quantitative data and an established formalism may have potential utility in the drug discovery industry if experimentation can be guided or even replaced by model-based simulations. Projects aimed at reenacting biological systems in silico, such as E-cell, The Virtual Cell, CyberCell, etc. (Normile 1999), combine the urge toward comprehensive characterization of Dimension 1 and detailed modeling. Practical and epistemological aspects of such modeling will be discussed under Dimension 5, in the context of the algorithmic complexity of living organisms.

In contrast, the other end of the spectrum that represents higher abstraction, which does not depend on quantitative data but rather on qualitative observations, is predominantly occupied by theoretical physicists and mathematicians who appreciate the necessity of simplification. Their operational goal is to find a minimal model that captures some essential generic aspect of a system's behavior. The long-term goal is to identify and understand generalizable, even universal principles rather than predict details of behavior of a particular instance. Therefore, their models typically contain "anonymous" genes and cells. They are interested in the potential behavioral repertoire rather than in predicting the actual behavior of a particular case. Historical examples include the Turing model of spatiotemporal patterns that can be produced by a set of partial differential equations (Turing 1952; Murray 1993; Meinhardt 1996), cellular automata-based models of cellular interaction (Ermentrout and Edelsteinkeshet 1993), or random Boolean genetic networks (Kauffman 1993). The distance between the two extremes of modeling approaches reflects the deep "scale space" discussed under Dimensions 2 and 3, and creates the need for middle-ground approaches. The practical necessity is obvious given the large number (tens of thousands) of different proteins active in cellular homeostasis and the desire to specify quantitatively the dynamic behavior of all the metabolic and signaling pathways. One example of such a compromise is the use of effective variables such as the "flux" in a specific metabolic pathway (Stephanopoulos et al. 1998), which allows precise descriptions at a fine scale to be converted into abstractions that are readily used at larger scales, much as statistical mechanics reduces to thermodynamics with ensemble averaging. Another abstraction is to focus on the stoichiometric or network constraints of biochemical reactions or on the structure of signaling pathways, while leaving out the physical kinetics (Papin et al. 2003; Ma'ayan et al. 2005). Without some "coarse-graining" it will be impossible to model the spatiotemporal distribution of all active proteins throughout a cell.

Not all mathematical approaches in biology serve the modeling and prediction of spatiotemporal behavior. Adjacent to the abstract end of the modeling spectrum are several branches of computational and theoretical biology, far beyond the "material world" of traditional systems biology, which after all is rooted in molecular biology. These areas deal with even more abstract but fundamental problems. They include, to mention only the most salient ones, the formal explanation of scaling laws in evolution, ecology, and physiology (Schmidt-Nielsen 1984; Brown et al. 2002); computational experimentation with simulated "artificial life" (Langton 1997); and the vast field of behavioral and cognitive biology, including models based on game theory (Imhof et al. 2005). If we view these endeavors as aiming at understanding the most nonmaterial, abstract, and emergent features of living systems, then their inclusion in systems biology may be justified by assigning them to this dimension of abstraction and emergence.

We conclude this section by noting that beyond technical and methodological issues, there is a fundamental limit to modeling, i.e., a useful description of a system by an abstract formalism, much as there is a limit to the compression of electronic files. This has to do with the concept of "computational irreducibility" discussed in the next section. For some aspects of sophistication of a system, there may be no theory to explain emergence, and no shortcut is possible, so that the detail-driven, exhaustive models aiming at a one-to-one reenacting of a specific instance of a real system would be the only way to predict behavior. One challenge, however, for such in silico step-by-step replay of the irreducible reality of a system is, in addition to the necessity to know all the details, the formidable computational cost, which we address in the last dimension of systems biology.

Dimension 5: "algorithmic" complexity

A central feature of complexity in a living organism is that it encodes and processes information. Technically, the term "algorithmic complexity" describes a measure for the computational resources needed to solve a computational problem. In a broader sense, we use it here both to refer to (1) the complexity of the "problems" that organisms compute and the physical medium they use for doing so, and (2) the complexity of the theoretical models required to simulate the organisms' computations. Viewing the organism as a computing machine opens many new perspectives: a living system is a physical system that computes its own development and homeostasis in response to external perturbations (Wolpert 1994). This is not an ad hoc analogy but has deeper implications that will become apparent, for it allows us to use concepts from computer science, and hence is distinct from the mathematical modeling used to address complexity.

One interesting question is the relationship between the physical implementation and the computation process. In von Neumann computers (which comprise most types of existing computer architectures) this is well understood; it is roughly captured in the dichotomy between hardware (transistors) and software (code). But how does biological computation exploit the physics of the living system? Computation by the organism is obviously a hybrid of digital and analog computing (Sarpeshkar 1998). While the DNA sequence (with the genetic code and transcriptional regulation as its programming language) is the most obvious embodiment of digital computing, the biochemistry of physiological and neuropsychological functions represents information processing that has analog components (molecule concentrations, membrane potentials). Digital and analog computing are intertwined. Gene regulatory networks in cells process information by transforming a continuous developmental signal, encoded by a hormone concentration as the input, into a gene expression pattern and associated discrete cellular phenotype as the output. By maintaining their distinct cell type identities despite harboring the same set of available instructions stored in the genome, cells also have an acquired memory of past events. At a higher level of organization, cell–cell communities form systems that also undertake computations, such as functional histological units, endocrine organs, or the immune system in which individual cells may act as discrete switches. The most prosaic system of multicellular computation is, of course, the brain and the peripheral sensory and neuromuscular system.

With our increasing knowledge of the physical medium that the living system uses for computing, one would then be interested in determining the computational efficiency, i.e., the use of energy, time, and space resources for computing. Analog computation costs less than digital computing chiefly because of the direct mapping of the device physics to operations and the lower communication overhead; however, the use of continuous variables leads to noise accumulation due to thermic fluctuations of molecular quantities (Sarpeshkar 1998; Kaern et al. 2005). Thus, a specific question of interest is: How does a living system use hybrid analog/digital computation to get the best of both and optimize computational efficiency, given the resource constraints and the challenge of performance degradation due to noise? In the hybrid architecture the digital components provide sets of discrete "attractor states" to which the analog signals can be reset in order to suppress noise accumulation (Cauwenberghs 1995). Such digital restoring states need not be embodied by an explicitly digital, molecular substrate, such as the nucleotide sequence. In fact, as discussed under Dimensions 3 and 4, the body is replete with nonlinear feedback circuit systems in molecular regulatory networks, electrochemical circuits or humoral inter-cell and inter-organ communications that produce stable oscillatory or stationary attractor states and all-or-none events, as found in biochemical switches (Ferrell and Machleder 1998; Tyson et al. 2003), action potentials, cell cycle progression, and cell fate determination (Huang 2005), which may act as the digital resetting attractor states (Barkai and Leibler 2000; Hasty et al. 2000). This is consistent with the use of transistors in digital computers—combinations of transistors, essentially analog devices, form digital flip-flops and from them logical (digital) AND, OR, and NOT gates.

It is important to note that the aspect of complexity in Dimension 5 is not easily captured by those formalisms of traditional mathematical modeling approaches which aim at predicting long-term dynamic behavior. Such modeling reduces the complexity by using a "shortcut" (in the form of a mathematical abstraction, such as sets of differential equations) to predict behavior. However, regarding the organism as a computing system leads us to the fundamental problem of "computational irreducibility" (Wolfram 2002). A system can be so sophisticated that the system used to describe and compute it (our brain, our mathematics, and our computers) is not sophisticated enough to "outrun" it. This may be the case with the attempt of a living organism (a human) to model a living organism: "If the [human] brain were so simple we could understand it, we would be so simple we couldn't" (Pugh 1977).

A consequence then is that mathematical modeling and general predictions may be impossible for some yet-to-be-identified aspects of organismal complexity. Instead, explicit simulation of every step, bit-by-bit, based on the knowledge of all the parts and a set of local interaction rules, would be theoretically the only way for computing such complex system behaviors at some reasonable resolution. However, epistemologically, such a one-to-one in silico replicate is not a model, but the thing itself. Therefore, it will be as difficult to interpret as reality. As explained above, there is no understanding without simplifying (Picht 1969). Thus, there may theoretically be an upper limit of sophistication only below which features of the living system are amenable to abstraction and mathematical modeling. The entities (subsystems, emerging properties) of a living system that are accessible to abstract modeling are represented by Dimension 4.

Computational irreducibility (Wolfram 2002) was discovered in the study of one of the most abstract types of models that biologists can fathom (or not): the cellular automaton (Ermentrout and Edelsteinkeshet 1993; Wolfram 2002). Thus, it is somewhat ironic that precisely the researchers pursuing the first dimension of complexity, who often dismiss abstract models intuitively or unknowingly, subliminally honor the principle of irreducibility by stressing the importance of exhaustive characterization of all possible molecular details. These biologists, habituated to "proximal explanations" and devoted to entireness of description rather than the study of entirety, in fact have always dreamed of an in silico bit-by-bit simulation of complex organisms using some "supercomputer" (Evans 2000). It was thought that with the knowledge of all the parts involved ("the words"), the rules of interaction ("grammar") (Bray 1997; Aebersold 2005) and with the aid of computers, we could overcome both the ontological failure (not knowing all the parts) and the epistemological failure (inability of the human brain to grasp the sophisticated interactions) that reductionism suffers from (Bray 1997). However, the idea of computational irreducibility touches upon a more subtle aspect of epistemology and implies that the sophistication of a complex system may reach into a regime where no abstraction into rules, no mathematical shortcut, is possible: there may be no grammar. Perhaps this is why "complex" is not simply "complicated."

In any case, the dream of exhaustively realistic models for bit-by-bit simulations among biologists exists—be it as an effort of burgeoning in silico biology in the spirit of our Dimension 1 of systems biology or (less likely) stimulated by the awareness of computational irreducibility. Such an endeavor faces practical challenges. For instance, even with the help of mathematical formalism as a shortcut, the electrical activity of the heart during ten seconds of fibrillation could easily require solving 10^{18} coupled differential equations (Cherry et al. 2000). (*N.B.*, Avogadro's number of differential equations may be defined as one Leibnitz, so 10 s of fibrillation corresponds to a micro-Leibnitz problem.) Multiprocessor supercomputers running for a month can execute a micromole of floating point operations, but in the cardiac case such computers may run several orders of magnitude slower than real time, such that modeling 10 s of fibrillation might require 1 exaFLOP/s×year. (*N.B.*, the Turing number may be defined as the ratio of the time required for a computation to time interval being simulated. A computer and program that could pass the Turing test for artificial intelligence might have a Turing number approaching unity.) Hence, it is easy to imagine bit-by-bit simulations of biological phenomena that are Leibnitz- and Turing-class problems that would take multiple lifetimes on the largest supercomputer imaginable. Thus, not without irony, here the loop closes: If shortcuts and coarse-graining in the spirit of mathematical modeling, as discussed under Dimension 4, are not satisfactory, and von Neumann computers are not powerful enough for a bit-by-bit reenacting of reality, then experimentation, again, becomes the method of choice for studying the complexity of life. It is in this context that we recognize organisms as massively parallel digital/analog computers. For instance, despite formidable efforts to create in silico hearts (Noble 2002), rabbit hearts will for a while to come still be used to "compute" the response of a drug-loaded heart to a stimulation or defibrillation protocol. DNA base-pairing is being explored for use in a new type of nanocomputers (Braich et al. 2002). Here, computational biology ends and biological computing begins. If an advanced computer and program could pass a full-fledged Turing test or fully model the details of a biological system, then the computation might be too complicated to understand. Now that may be beyond systems biology.

Conclusion

We did not intend to define systems biology here, nor do we claim this review to be comprehensive. We have focused on biology as an analytic science that seeks to understand the living organism as a complex entity, while leaving out discussions of areas of biology that represent synthetic science aimed at building useful systems, such as genetic, cell, and tissue engineering. (Of course, a synthetic approach also can contribute to understanding a class of a system.) A rigorous definition of systems biology would also have to identify what is really new and different from classical biophysics and mathematical or computational biology, and in particular, what distinguishes it from the previous waves of "anti-reductionism" in the past century that explicitly addressed the phenomenon of a "system" as such, including cybernetics (Ashby 1964; Wiener 1965), general systems theory (von Bertalanffy 1969), and the more recent complex systems sciences (Waldrop 1992; Bar-Yam 1997), whose decades-old manifestos sometimes read like those of the websites of modern initiatives in systems biology. Perhaps its novelty lies in the fact that systems biology is the first anti-reductionist movement that is broadly endorsed by mainstream experimental biologists and that emphasizes active experimentation in iteration with theory, rather than relying upon thinking about experimental observations made by others. There were times in biology when we had "theories but no data," then molecular biology brought an era of "data but no theories"—perhaps now we enter the promising age of "theories with data" (and vice versa).

Yet caution is warranted. Because of its apparent claim to encompass all, systems biology may appear to be a biological discipline of everything. As we all know, a discipline of everything needs no name, and a theory that explains everything explains nothing. But we also need to avoid too narrow a view. Many initiatives in systems biology tend to describe their effort in terms of logistic novelty, namely, the multidisciplinary approach that integrates quantitative and computational methods with experimental biology, without embracing the concepts and addressing the challenges of complexity. Understanding "complexity" may not require a fundamentally novel theory at all (Horgan 1995), but certainly, it will entail adopting a more encompassing and pluralistic mind-set than the one that was sufficient for the characterization of proteins and pathways of the past decades. Since we think that the term "systems biology" is justified and useful, we have here tried to organize this new discipline into five dimensions, which reflect both the various approaches embodied by existing academic disciplines as well as the various aspects of complexity of the organism. The five dimensions should serve as an aid to systemize the immense diversity of the approaches and research questions. They should not be understood as five mutually exclusive directions or sub-disciplines of systems biology, but as a characterization of approaches that we think represent the most natural components of both systems biology as a scientific endeavor and living organisms as complex systems. Thus, while not independent, the five idealized axes can be treated as quasi-orthogonal, so that any given research in reality will be a combination of various contributions of all five dimensions. The true challenge for systems biology will be to bridge the diverse cultural and mental habits of scientists working along these different axes within one system, one project, and one question. It is in this context that we can view systems biology as an emergent discipline of biology, in every sense of that word.

Acknowledgements. We are deeply indebted to the many colleagues and friends who, through their publications, conversations, and presentations, have helped educate us about many aspects of systems biology and have critiqued our views. We apologize to those whose work is not explicitly mentioned. A list of all original references would have been irresponsibly long. Our citations also do not meet the standard of a disciplined historian of science. Rather than consistently reflecting the originators of an idea, the citations were chosen to be most practical and instructive for readers who seek further reading on a subject. We are indebted to Allison Price and Don Berry for their unflagging editorial assistance with this manuscript.

References

Aebersold R (2005) Molecular systems biology: a new journal for a new biology? Mol Sys Biol 1:msb4100009 E1–E2
Aebersold R, Hood LE, Watts JD (2000) Equipping scientists for the new biology. Nat Biotechnol 18:359
Albert R, Jeong H, Barabasi AL (2000) Error and attack tolerance of complex networks. Nature 406:378–382
Alon U (2003) Biological networks: the tinkerer as an engineer. Science 301:1866–1867
Alon U, Surette MG, Barkai N, Leibler S (1999) Robustness in bacterial chemotaxis. Nature 397:168–171
Anderson PW (1972) More is different: broken symmetry and the nature of the hierarchical structure of science. Science 177:393–396
Ashby WR (1964) An introduction to cybernetics. Routledge, Kegan and Paul, London
Autumn K, Ryan MJ, Wake DB (2002) Integrating historical and mechanistic biology enhances the study of adaptation. Q Rev Biol 77:383–408
Balaban NQ, Merrin J, Chait R, Kowalik L, Leibler S (2004) Bacterial persistence as a phenotypic switch. Science 305:1622–1625
Ball P (2001) The self-made tapestry: pattern formation in nature. Oxford University Press, Oxford
Bar-Yam Y (1997) Dynamics of complex systems. Perseus Books, Reading
Bar-Yam Y (2004) Multiscale variety in complex systems. Complexity 9:37–45
Barabasi AL, Albert R (1999) Emergence of scaling in random networks. Science 286:509–512
Barabasi AL, Oltvai ZN (2004) Network biology: understanding the cell's functional organization. Nat Rev Genet 5:101–113
Barkai N, Leibler S (2000) Circadian clocks limited by noise. Nature 403:267–268
Blackett PMS (1963) Memories of Rutherford. In: Birks JB (ed) Rutherford at Manchester. Benjamin, New York, p 108
Bassingthwaighte JB, Liebovitch LS, West BJ (1994) Fractal physiology. Oxford University Press, New York
Braich RS, Chelyapov N, Johnson C, Rothemund PWK, Adleman L (2002) Solution of a 20-variable 3-SAT problem on a DNA computer. Science 296:499–502
Bray D (1997) Reductionism for biochemists: how to survive the protein jungle. Trends Biochem Sci 22:325–326
Brent R (2000) Genomic biology. Cell 100:169–183
Brown JH, Gupta VK, Li BL, Milne BT, Restrepo C, West GB (2002) The fractal nature of nature: power laws, ecological complexity and biodiversity. Philos Trans R Soc Lond B Biol Sci 357:619–626
Carlson JM, Doyle J (2002) Complexity and robustness. Proc Natl Acad Sci USA 99:2538–2545
Cauwenberghs G (1995) A micropower CMOS algorithmic A/D/A converter. IEEE T Circuits Syst I 42:913–919
Cherry EM, Greenside HS, Henriquez CS (2000) A space-time adaptive method for simulating complex cardiac dynamics. Phys Rev Lett 84:1343–1346
Corning PA (2002) The re-emergence of "emergence": a venerable concept in search of a theory. Complexity 7:18–30
Cross SS (1997) Fractals in pathology. J Pathol 182:1–8
Cummings DAT, Irizarry RA, Huang NE, Endy TP, Nisalak A, Ungchusak K, Burke DS (2004) Travelling waves in the occurrence of dengue haemorrhagic fever in Thailand. Nature 427:344–347
D'haeseleer P, Liang SD, Somogyi R (2000) Genetic network inference: from co-expression clustering to reverse engineering. Bioinformatics 16:707–726

Davidson EH, Rast JP, Oliveri P, Ransick A, Calestani C, Yuh CH, Minokawa T, Amore G, Hinman V, Arenas-Mena C, Otim O, Brown CT, Livi CB, Lee PY, Revilla R, Rust AG, Pan ZJ, Schilstra MJ, Clarke PJC, Arnone MI, Rowen L, Cameron RA, McClay DR, Hood L, Bolouri H (2002) A genomic regulatory network for development. Science 295:1669–1678

Dawkins R (1996) The blind watchmaker: why the evidence of evolution reveals a universe without design. Norton, New York

Dretske FI (2000) Perception, knowledge, and belief: selected essays. Cambridge University Press, Cambridge

Eklund SE, Taylor D, Kozlov E, Prokop A, Cliffel DE (2004) A microphysiometer for simultaneous measurement of changes in extracellular glucose, lactate, oxygen, and acidification rate. Anal Chem 76:519–527

Elsasser WM (1998) Reflections on a theory of organisms: holism in biology. Johns Hopkins University Press, Baltimore

Endy D, Brent R (2001) Modelling cellular behaviour. Nature 409:391–395

Enver T, Heyworth CM, Dexter TM (1998) Do stem cells play dice? Blood 92:348–351

Ermentrout GB, Edelsteinkeshet L (1993) Cellular automata approaches to biological modeling. J Theor Biol 160:97–133

Evans GA (2000) Designer science and the "omic" revolution. Nat Biotechnol 18:127

Ferrell JE, Machleder EM (1998) The biochemical basis of an all-or-none cell fate switch in Xenopus oocytes. Science 280:895–898

Fitzhugh R (1961) Impulses and physiological states in theoretical models of nerve membrane. Biophys J 1:445–466

Freitas RA Jr (2002) The future of nanofabrication and molecular scale devices in nanomedicine. Stud Health Technol Inform 80:45–59

Gardner TS, di Bernardo D, Lorenz D, Collins JJ (2003) Inferring genetic networks and identifying compound mode of action via expression profiling. Science 301:102–105

Ge H, Walhout AJM, Vidal M (2003) Integrating 'omic' information: a bridge between genomics and systems biology. Trends Genet 19:551–560

Gell-Mann M (1995) The quark and the jaguar: adventures in the simple and the complex. WH Freeman, New York

Glass L (2001) Synchronization and rhythmic processes in physiology. Nature 410:277–284

Gleick J (1988) Chaos: making a new science. Penguin, New York

Goldberger AL, Amaral LAN, Hausdorff JM, Ivanov PC, Peng CK, Stanley HE (2002) Fractal dynamics in physiology: alterations with disease and aging. Proc Natl Acad Sci USA 99:2466–2472

Goodwin BC (2001) How the leopard changed its spots: the evolution of complexity. Princeton University Press, Princeton

Goodwin BC, Kauffman S, Murray JD (1993) Is morphogenesis an intrinsically robust process? J Theor Biol 163:135–144

Gould SJ (2002) The structure of evolutionary theory. Belknap Press of Harvard University Press, Cambridge

Gould SJ, Lewontin RC (1979) The spandrels of San Marco and the Panglossian paradigm: a critique of the adaptationist programme. Proc R Soc Lond B Biol Sci 205:581–598

Graveley BR (2001) Alternative splicing: increasing diversity in the proteomic world. Trends Genet 17:100–107

Gray RA, Jalife J, Panfilov AV, Baxter WT, Cabo C, Davidenko JM, Pertsov AM (1995) Nonstationary vortexlike reentrant activity as a mechanism of polymorphic ventricular tachycardia in the isolated rabbit heart. Circulation 91:2454–2469

Hall D, Minton AP (2003) Macromolecular crowding: qualitative and semiquantitative successes, quantitative challenges. Biochim Biophys Acta 1649:127–139

Hasty J, Pradines J, Dolnik M, Collins JJ (2000) Noise-based switches and amplifiers for gene expression. Proc Natl Acad Sci USA 97:2075–2080

Horgan J (1995) From complexity to perplexity. Sci Am 272:104–109

Huang S (2004) Back to the biology in systems biology: what can we learn from biomolecular networks? Brief Funct Genomic Proteomic 2:279–297

Huang S (2005) Multistability and multicellularity: cell fates as high-dimensional attractors of gene regulatory networks. In: Kriete A,Eils R (eds) Computational systems biology. Elsevier, Amsterdam, pp 293–326

Huang S, Eichler G, Bar-Yam Y, Ingber DE (2005) Cell fates as high-dimensional attractor states of a complex gene regulatory network. Phys Rev Lett 94:128701

Hume DA (2000) Probability in transcriptional regulation and its implications for leukocyte differentiation and inducible gene expression. Blood 96:2323–2328

Ideker T, Galitski T, Hood L (2001) A new approach to decoding life: systems biology. Annu Rev Genomics Hum Genet 2:343–372

Imhof LA, Fudenberg D, Nowak MA (2005) Evolutionary cycles of cooperation and defection. Proc Natl Acad Sci USA 102:10797–10800
Ingber DE (1998) The architecture of life. Sci Am 278:48–57
Ingber DE (2003) Tensegrity I. Cell structure and hierarchical systems biology. J Cell Sci 116:1157–1173
Ivanov PC, Amaral LAN, Goldberger AL, Havlin S, Rosenblum MG, Struzik ZR, Stanley HE (1999) Multifractality in human heartbeat dynamics. Nature 399:461–465
Kaern M, Elston TC, Blake WJ, Collins JJ (2005) Stochasticity in gene expression: from theories to phenotypes. Nat Rev Genet 6:451–464
Kauffman SA (1993) The origins of order: self-organization and selection in evolution. Oxford University Press, New York
Kirschner M, Gerhart J, Mitchison T (2000) Molecular "vitalism". Cell 100:79–88
Kitano H, Oda K, Kimura T, Matsuoka Y, Csete M, Doyle J, Muramatsu M (2004) Metabolic syndrome and robustness tradeoffs. Diabetes 53 [Suppl 3]:S6–S15
Klevecz RR, Bolen J, Forrest G, Murray DB (2004) A genomewide oscillation in transcription gates DNA replication and cell cycle. Proc Natl Acad Sci USA 101:1200–1205
Kocer A, Walko M, Meijberg W, Feringa BL (2005) A light-actuated nanovalve derived from a channel protein. Science 309:755–758
Lahav G, Rosenfeld N, Sigal A, Geva-Zatorsky N, Levine AJ, Elowitz MB, Alon U (2004) Dynamics of the p53-Mdm2 feedback loop in individual cells. Nat Genet 36:147–150
Langton CG (1997) Artificial life: an overview. MIT Press, Cambridge
Laurent M, Kellershohn N (1999) Multistability: a major means of differentiation and evolution in biological systems. Trends Biochem Sci 24:418–422
Le TT, Harlepp S, Guet CC, Dittmar K, Emonet T, Pan T, Cluzel P (2005) Real-time RNA profiling within a single bacterium. Proc Natl Acad Sci USA 102:9160–9164
Levsky JM, Singer RH (2003) Gene expression and the myth of the average cell. Trends Cell Biol 13:4–6
Levsky JM, Shenoy SM, Pezo RC, Singer RH (2002) Single-cell gene expression profiling. Science 297:836–840
Lewontin RC (2001) The triple helix: gene, organism, and environment. Harvard University Press, Cambridge
Lowrey PL, Takahashi JS (2004) Mammalian circadian biology: elucidating genome-wide levels of temporal organization. Annu Rev Genomics Hum Genet 5:407–441
Lu JZ, Rosenzweig Z (2000) Nanoscale fluorescent sensors for intracellular analysis. Fresenius J Anal Chem 366:569–575
Ma'ayan A, Jenkins SL, Neves S, Hasseldine A, Grace E, Dubin-Thaler B, Eungdamrong NJ, Weng G, Ram PT, Rice JJ, Kershenbaum A, Stolovitzky GA, Blitzer RD, Iyengar R (2005) Formation of regulatory patterns during signal propagation in a mammalian cellular network. Science 309:1078–1083
Mackey MC, Glass L (1977) Oscillation and chaos in physiological control systems. Science 197:287–288
Mandelbrot BB (1982) The fractal geometry of nature. WH Freeman, San Francisco
Mangan S, Alon U (2003) Structure and function of the feed-forward loop network motif. Proc Natl Acad Sci USA 100:11980–11985
Marcotte EM (2001) The path not taken. Nat Biotechnol 19:626–627
Marsh BJ, Mastronarde DN, Buttle KF, Howell KE, McIntosh JR (2001) Organellar relationships in the Golgi region of the pancreatic beta cell line, HIT-T15, visualized by high resolution electron tomography. Proc Natl Acad Sci USA 98:2399–2406
Medina M (2005) Genomes, phylogeny, and evolutionary systems biology. Proc Natl Acad Sci USA 102 [Suppl 1]:6630–6635
Meinhardt H (1996) Models of biological pattern formation: common mechanism in plant and animal development. Int J Dev Biol 40:123–134
Milo R, Shen-Orr S, Itzkovitz S, Kashtan N, Chklovskii D, Alon U (2002) Network motifs: simple building blocks of complex networks. Science 298:824–827
Morange M (2001) The misunderstood gene. Harvard University Press, Cambridge
Murray JD (1993) Mathematical biology. Springer-Verlag, Heidelberg, Berlin, New York
Nguyen TD, Tseng HR, Celestre PC, Flood AH, Liu Y, Stoddart JF, Zink JI (2005) A reversible molecular valve. Proc Natl Acad Sci USA 102:10029–10034
Nicolis G, Prigogine I (1989) Exploring complexity: an introduction. WH Freeman, New York
Ninfa AJ, Mayo AE (2004) Hysteresis vs. graded responses: the connections make all the difference. Science's STKE 2004:e20
Noble D (2002) Modeling the heart—from genes to cells to the whole organ. Science 295:1678–1682
Normile D (1999) Building working cells 'in silico'. Science 284:80–81
Okubo A (1986) Dynamical aspects of animal grouping: swarms, schools, flocks, and herds. Adv Biophys 22:1–94

Ozbudak EM, Thattai M, Lim HN, Shraiman BI, van Oudenaarden A (2004) Multistability in the lactose utilization network of Escherichia coli. Nature 427:737–740
Palsson E, Cox EC (1996) Origin and evolution of circular waves and spirals in Dictyostelium discoideum territories. Proc Natl Acad Sci USA 93:1151–1155
Papin JA, Price ND, Wiback SJ, Fell DA, Palsson BO (2003) Metabolic pathways in the post-genome era. Trends Biochem Sci 28:250–258
Pattee HH (1973) Hierarchy theory: the challenge of complex systems. G Braziller, New York
Picht G (1969) Mut zur Utopie: die grossen Zukunftsaufgaben; zwölf Vorträge. R Piper, München
Pugh GE (1977) The biological origin of human values. Basic Books, New York
Reik W, Dean W (2002) Back to the beginning. Nature 420:127
Rocheleau JV, Walker GM, Head WS, McGuinness OP, Piston DW (2004) Microfluidic glucose stimulation reveals limited coordination of intracellular Ca2+ activity oscillations in pancreatic islets. Proc Natl Acad Sci USA 101:12899–12903
Rose SPR (2003) Lifelines: life beyond the gene. Oxford University Press, Oxford
Rubin H (1990) On the nature of enduring modifications induced in cells and organisms. Am J Physiol 258:L19–L24
Sachs K, Perez O, Pe'er D, Lauffenburger DA, Nolan GP (2005) Causal protein-signaling networks derived from multiparameter single-cell data. Science 308:523–529
Sarpeshkar R (1998) Analog versus digital: extrapolating from electronics to neurobiology. Neural Comput 10:1601–1638
Schmidt-Nielsen K (1984) Scaling: why is animal size so important. Cambridge University Press, New York
Simon SM, Llinas RR (1985) Compartmentalization of the submembrane calcium activity during calcium influx and its significance in transmitter release. Biophys J 48:485–498
Sohrmann M, Peter M (2003) Polarizing without a C(l)ue. Trends Cell Biol 13:526–533
Southan C (2004) Has the yo-yo stopped? An assessment of human protein-coding gene number. Proteomics 4:1712–1726
Springel V, White SDM, Jenkins A, Frenk CS, Yoshida N, Gao L, Navarro J, Thacker R, Croton D, Helly J, Peacock JA, Cole S, Thomas P, Couchman H, Evrard A, Colberg J, Pearce F (2005) Simulations of the formation, evolution and clustering of galaxies and quasars. Nature 435:629–636
Spudich JL, Koshland DE (1976) Non-genetic individuality: chance in the single cell. Nature 262:467–471
Stelling J, Sauer U, Szallasi Z, Doyle FJ, Doyle J (2004) Robustness of cellular functions. Cell 118:675–685
Stephanopoulos GN, Aristidou AA, Nielsen J (1998) Metabolic engineering: principles and methodologies. Academic Press, San Diego
Strogatz SH (2001) Exploring complex networks. Nature 410:268–276
Strohman RC (1997) The coming Kuhnian revolution in biology. Nat Biotechnol 15:194–200
Strohman RC (2000) Organization becomes cause in the matter. Nat Biotechnol 18:575–576
Stuart JM, Segal E, Koller D, Kim SK (2003) A gene-coexpression network for global discovery of conserved genetic modules. Science 302:249–255
Takahashi K, Arjunan SNV, Tomita M (2005) Space in systems biology of signaling pathways—towards intracellular molecular crowding in silico. FEBS Lett 579:1783–1788
Tinbergen N (1952) Derived activities: their causation, biological significance, origin, and emancipation during evolution. Q Rev Biol 27:1–32
Turing AM (1952) The chemical basis of morphogenesis. Philos Trans R Soc Lond B Biol Sci 237:37–72
Tyson JJ, Chen KC, Novak B (2003) Sniffers, buzzers, toggles and blinkers: dynamics of regulatory and signaling pathways in the cell. Curr Opin Cell Biol 15:221–231
von Bertalanffy L (1969) General system theory; foundations, development, applications. G Braziller, New York
Waddington CH (1956) Principles of embryology. Allen and Unwin, London
Waldrop MM (1992) Complexity: the emerging science at the edge of order and chaos. Simon and Schuster, New York
Webster G, Goodwin BC (1984) A structuralist approach to morphology. Riv Biol 77:503–531
Whitesides GM (2003) The 'right' size in nanobiotechnology. Nat Biotechnol 21:1161–1165
Whitfield ML, Sherlock G, Saldanha AJ, Murray JI, Ball CA, Alexander KE, Matese JC, Perou CM, Hurt MM, Brown PO, Botstein D (2002) Identification of genes periodically expressed in the human cell cycle and their expression in tumors. Mol Biol Cell 13:1977–2000
Wiener N (1965) Cybernetics: or control and communication in the animal and the machine. MIT Press, Cambridge
Wikswo JP, Prokop A, Baudenbacher F, Cliffel D, Csukas B, Velkovsky M (2006) The engineering challenges of BioNEMS: the integration of microfluidics, and micro- and nanodevices, models, and external control for systems biology. IEE Proceedings Nanobiotechnology (in press)

Wilders R, Jongsma HJ (1993) Beating irregularity of single pacemaker cells isolated from the rabbit sinoatrial node. Biophys J 65:2601–2613

Wilson EO (1995) Naturalist. Warner Books, New York

Wolfe MF, Goldberg R (2000) Rube Goldberg: inventions. Simon and Schuster, New York

Wolfram S (2002) A new kind of science. Wolfram Media, Champaign

Wolpert L (1994) Do we understand development? Science 266:571–572

Xiong W, Ferrell JE (2003) A positive-feedback-based bistable 'memory module' that governs a cell fate decision. Nature 426:460–465

Yamamoto T, Nakahata Y, Soma H, Akashi M, Mamine T, Takumi T (2004) Transcriptional oscillation of canonical clock genes in mouse peripheral tissues. BMC Mol Biol 5:18

Yi TM, Huang Y, Simon MI, Doyle J (2000) Robust perfect adaptation in bacterial chemotaxis through integral feedback control. Proc Natl Acad Sci USA 97:4649–4653

Rev Physiol Biochem Pharmacol (2006)
DOI 10.1007/112_0603

S. C. Sinha · S. R. Sprang

Structures, mechanism, regulation and evolution of class III nucleotidyl cyclases

Published online: 12 September 2006

Abstract Cyclic 3′,5′-guanylyl and adenylyl nucleotides function as second messengers in eukaryotic signal transduction pathways and as sensory transducers in prokaryotes. The nucleotidyl cyclases (NCs) that catalyze the synthesis of these molecules comprise several evolutionarily distinct groups, of which class III is the largest. The domain structures of prokaryotic and eukaryotic class III NCs are diverse, including a variety of regulatory and transmembrane modules. Yet all members of this family contain one or two catalytic domains, characterized by an evolutionarily ancient topological motif (βααββαβ) that is preserved in several other enzymes that catalyze the nucleophilic attack of a 3′-hydroxyl upon a 5′ nucleotide phosphate. Two dyad-related catalytic domains compose one catalytic unit, with the catalytic sites formed at the domain interface. The catalytic domains of mononucleotidyl cyclases (MNCs) and diguanylate cyclases (DGCs) are called cyclase homology domains (CHDs) and GGDEF domains, respectively. Prokaryotic NCs usually contain only one catalytic domain and are catalytically active as intermolecular homodimers. The different modes of dimerization in class III NCs probably evolved concurrently with their mode of binding substrate. The catalytic mechanism of GGDEF domain homodimers is not completely understood, but they are expected to have a single active site with each subunit contributing equivalent determinants to bind one GTP molecule or half a c-diGMP molecule. CHD dimers have two potential dyad-related active sites, with both CHDs contributing determinants to each site. Homodimeric class III MNCs have two equivalent catalytic sites, although such enzymes may show half-of-sites reactivity. Eukaryotic class III MNCs often contain two divergent CHDs, with only one catalytically competent site. All CHDs appear to use a common catalytic mechanism, which requires the participation of two magnesium

S. C. Sinha (✉)
University of Texas Southwestern Medical Center at Dallas, Division of Infectious Diseases, Department of Internal Medicine,
5323 Harry Hines Blvd., Dallas, 75390-9113 TX, USA
e-mail: Sangita.Sinha@UTSouthwestern.edu · Tel.: +1-214-6487240 · Fax: +1-214-6480248

S. R. Sprang
University of Texas Southwestern Medical Center, Department of Biochemistry,
6001 Forest Park Rd., Dallas, 75390-8816 TX, USA

or manganese ions for binding polyphosphate groups and nucleophile activation. In contrast, mechanisms for purine recognition and specificity are more diverse. Class III NCs are subject to regulation by small molecule effectors, endogenous domains, or exogenous protein partners. Many of these regulators act by altering the interface of the catalytic domains and therefore the integrity of the catalytic site(s). This review focuses on both conserved and divergent mechanisms of class III NC function and regulation.

Introduction

Small-molecule second messengers such as inositol triphosphate, diacylglycerol, guanosine-5′-triphosphate-3′-diphosphate, Ca^{2+}, and the cyclic nucleotides (cNMPs), adenosine 3′-5′ cyclic monophosphate (cAMP), guanosine 3′-5′ cyclic monophosphate (cGMP) and bis-(3′-5′)-cyclic di-guanosine monophosphate (c-diGMP), are key components of intracellular signal transduction pathways in all cellular organisms. The first small-molecule second messenger to be identified was cAMP, while c-diGMP is the most recently discovered (Barzu and Danchin 1994; McCue et al. 2000; Romling et al. 2005). While cAMP has been found in most cellular organisms except plants, cGMP has been found chiefly in eukaryotes, and c-diGMP appears to be ubiquitous among prokaryotes (Cooper 2003; Linder and Schultz 2003; Romling et al. 2005; Shenoy and Visweswariah 2004). The cNMPs regulate cellular functions such as gene expression, metabolism, and ion flux in organisms ranging from prokaryotes to mammals (Romling et al. 2005; Taussig and Zimmermann 1998). In metazoans, cyclic nucleotides mediate processes such as cell growth, differentiation during embryogenesis, cardiac contractility, transmission of nerve impulses, learning and memory, and blood glucose homeostasis (Krupinski and Cali 1998; Smit and Iyengar 1998; Taussig and Zimmermann 1998), while in unicellular organisms they have been shown to regulate motility, chemotaxis, pathogenicity, response to osmotic stress and environmental acidification, cell-to-cell-communication, biofilm formation, and multicellular behavior (Kimura et al. 1997; Ladant and Ullmann 1999; Leppla 1982; Romling et al. 2005; Süsstrunk et al. 1998; Yahr et al. 1998). Intracellular levels of cNMPs are primarily regulated by their rate of synthesis, although localization or subcellular compartmentalization, as well as their rate of degradation by enzymes called phosphodiesterases, also plays an important role (Cooper 2003; Hanoune and Defer 2001). In vivo, cAMP, cGMP, and c-diGMP are synthesized by nucleotidyl cyclases (NCs): adenylyl cyclases (ACs), guanylyl cyclases (GCs), and diguanylate cyclases (DGCs), respectively. A comprehensive review by Barzu and Danchin grouped mononucleotidyl cyclases (MNCs) that had been identified into three classes on the basis of amino acid sequence, and summarized the biochemical information available for each class (Barzu and Danchin 1994). Subsequently, reviews of NCs have chiefly focused on the most universal of these three classes, the class III NCs. In the last few years, the structures of the catalytic domains of various class III NCs have been solved. This review will primarily focus on knowledge obtained from these structures in the context of available biochemical information.

Classification

All NCs that have been identified are grouped into five nonhomologous classes based on sequence similarity (Barzu and Danchin 1994; Cotta et al. 1998; Linder and Schultz 2003;

Sismeiro et al. 1998). Each of the five classes of cyclases is expected to have a different structural fold and mode of activity, suggesting that each class probably arose from different ancestral enzymes. Despite these key differences, all NCs use nucleoside triphosphates (NTPs) as substrate and catalyze identical chemical reactions, and consequently have convergently evolved some common features for binding and catalysis. Except for the class III NCs, enzymes belonging to the other classes appear to be MNCs that are strictly specific for ATP.

Class I NCs are ACs that have been only found in gram-negative, enteric, γ-proteobacteria such as *Escherichia, Yersinia, Haemophilus* and *Pseudomonas*. The class I ACs from these bacteria are approximately 850-amino-acid proteins that are thought to contain two domains: an N-terminal, catalytic domain of approximately 450 amino acids and a C-terminal, regulatory domain of approximately 500 amino acids (Barzu and Danchin 1994; Holland et al. 1988; Reddy et al. 1995a). The primary structure of all ACs that have been grouped in class I, which share between 50% and 99% sequence identity, is well conserved. The class I ACs were the first NCs to be identified, yet have not been extensively characterized and there are no representative structures. Therefore, residues important for the activity of these ACs have not yet been identified.

Class II NCs are AC exotoxins secreted by pathogenic prokaryotes such as *Bacillus anthracis, Bordetella pertussis*, and *Pseudomonas aeroginosa* (Barzu and Danchin 1994; Ladant and Ullmann 1999; Leppla 1982; Yahr et al. 1998). These ACs differ considerably in length, with the *B. anthracis, B. pertussis*, and *P. aeroginosa* enzymes comprising 800, 1,706, and 378 amino acids respectively. The catalytic fragments of each of these class II ACs are approximately 360 amino acids in length and share less than 30% sequence identity with each other. The architecture of the class II ACs from these three organisms diverges greatly, with the conserved catalytic fragment attached to a variety of domains thought to play a role in regulation. These cyclases are activated by different host cell factors and are inactive until they are injected into the host cell. Structures of the catalytic domains of two class II ACs, edema factor (EF) from *B. anthracis* and, more recently, CyaA from *B. pertussis*, have been determined (Drum et al. 2002; Guo et al. 2004, 2005; Shen et al. 2005).

Class III constitutes the largest and most diverse family of NCs. It includes all DGCs, GCs, and eukaryotic ACs identified to date, as well as ACs from many prokaryotes. The catalytic domains of class III NCs are approximately 180 amino acids in length. The catalytic domains of the class III MNCs, the ACs and GCs, are often called cyclase homology domains (CHDs), while those of the DGCs are called GGDEF domains, based on the presence of a highly conserved Gly-Gly-Asp-Glu-Phe sequence motif. Notably, the central Gly-Asp couple of this motif is a very highly conserved sequence feature of the CHDs as well. Both CHDs and GGDEF domains are highly divergent and therefore difficult to identify solely from sequence. In fact, only four residues, an aspartate, the Gly-Asp motif described above, and another glycine are well-conserved among all class III NCs (Fig. 1). The biochemical significance of the numerous variations among the catalytic domains of class III NCs, such as differences in residues implicated in catalysis and insertions in the core fold that may be responsible for varying modes of regulation, has yet to be completely understood.

The evolutionary relationship between CHDs and GGDEF domains has been previously investigated (Pei and Grishin 2001). It was noted that outliers among the GGDEF domains sometimes share higher sequence identity with some CHDs, than with other GGDEF domains. The small size of these domains combined with their divergent sequences prevented construction of a reliable phylogenetic tree, yet it was clear that GGDEF domains and CHDs belong to separate groups. GGDEF domains may be further subdivided into two clusters, one comprising of DGCs from eubacteria, and the other including DGCs from archaebacteria as

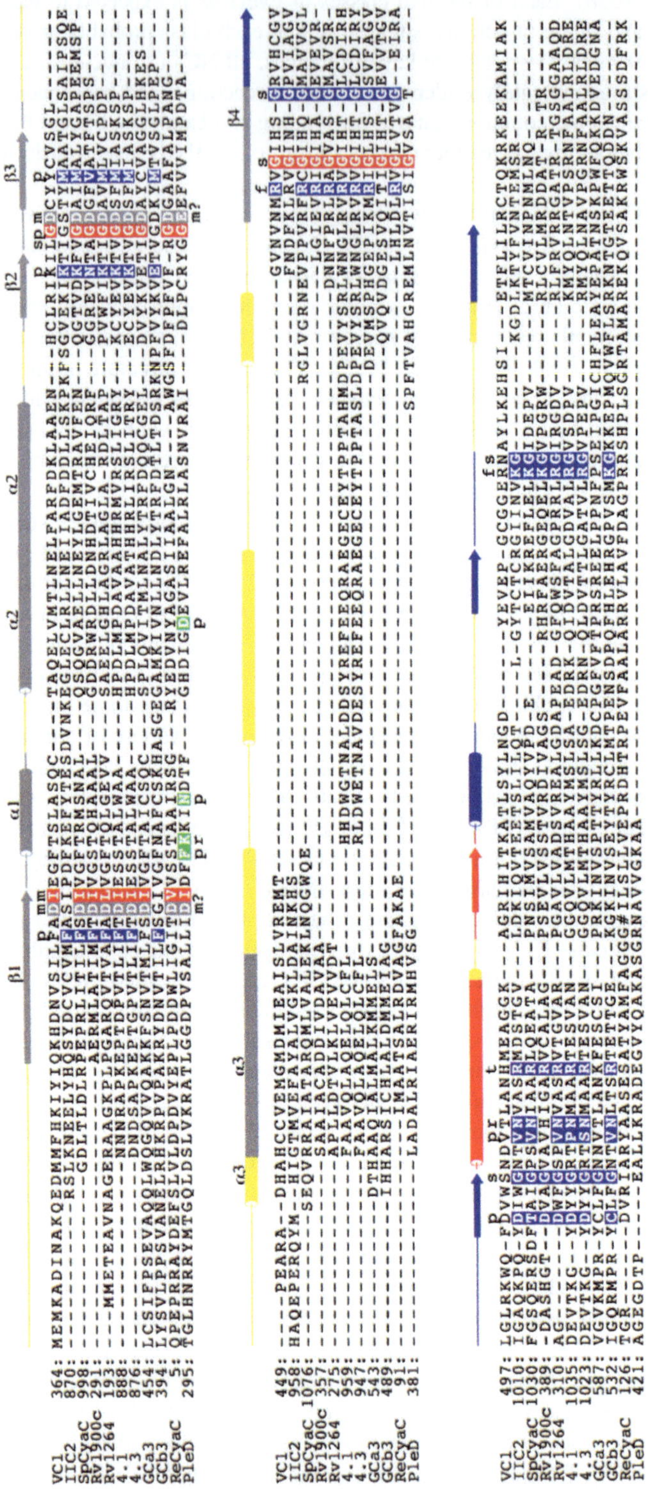

◀ **Fig. 1** Structure-based sequence alignments of catalytic domains of class III NCs. Biological sources and either PDB codes for class III NCs of known structure or accession numbers for other class III NC sequences are: *VC1* Canine AC VC_1 (1CJK_A), *IIC2* Rat AC IIC_2 (1CJK_B), *SpCyaC Spirulina platensis* CyaC (1WC6), *Rv1900c* and *Rv1264 Mycobacterium tuberculosis* Rv1900c (1YBU) and Rv1264 (1Y11) respectively; *4.1* and *4.3 Trypanosoma brucei* GRESAGs 4.1 (1FX2) and 4.3 (1FX4) respectively, *GCa3* and *GCb3* Canine soluble GC1 subunits α3 (gi:65301169) and β3 (gi:65294809) respectively, *ReCyaC Rhizobium etli* CyaC (gi: 20530211), *PleD Caulobacter crescentus* (1W25). The first residue of each sequence in each alignment bloc is indicated. The "#" symbol indicates residues 151–191 of ReCyaC which are omitted from the alignment. Consensus secondary structure elements deduced from class III NCs of known structure are shown *above the alignment* and are numbered according to the core class III NC fold, and not in context of the individual full-length proteins. *Bold white letters* highlighted against *colored backgrounds* indicate conserved residues important for the activity of class III NCs. The role of these conserved residues is indicated *above* and *below* the alignment for CHDs and GGDEF domains, respectively, by: *s* structural, *m* metal-coordinating, *f* phosphate-bonding, *r* ribose-bonding, *p* purine-interacting. Secondary structure elements and conserved residues common to both class III NCs and type I DNA polymerases are highlighted in *gray*, those common to all class III NCs are highlighted in *red*, those conserved only among either the CHDs or GGDEF domains are highlighted in *blue* or *green*, respectively, and where possible, features unique to individual enzymes are shown in *yellow*. This figure was made using the program ALSCRIPT (Barton 1993)

well as some eubacteria. Recently, the first structure of a GGDEF domain, PleD, a DGC from *Caulobacter crescentus* was solved, and conclusively demonstrated that this domain belongs to the class III NCs (Table 1; Chan et al. 2004).

As for the GGDEF domains, some divergent CHDs share higher sequence identity with some GGDEF domains than with the more closely related members within the CHD family. Linder and Schultz have subdivided CHDs into subclasses IIIa–d based on variations in residues required for catalysis as well as the length of a β-ribbon located between a glycine and an aspartate that are conserved among most CHDs (Fig. 1; Linder and Schultz 2003). McCue et al. have subdivided mycobacterial CHDs into classes I–V according to the statistical occurrence of amino acids at each residue position (McCue et al. 2000). The functional, structural, and evolutionary implications of these subdivisions have yet to be completely explored. Available CHD structures include those of the C_1 domain of type V mammalian AC (VC_1) and the C_2 domain of type II mammalian AC (IIC_2), trypanosomal GRESAG 4.1 and 4.3, *Spirulina platensis* CyaC, and *Mycobacterium tuberculosis* Rv1900c, Rv1264, and Rv1625c (Table 1; Bieger and Essen 2001; Ketkar et al. 2006; Mou et al. 2005; Sinha et al. 2005; Steegborn et al. 2005a, b; Tesmer et al. 1997, 1999, 2000; Tews et al. 2005; Zhang et al. 1997).

The architecture of class III NCs diverges greatly. However, mammalian ACs, which have been the focus of extensive biochemical and structural analyses, are fairly well conserved. Nine of ten mammalian ACs that have been identified are membrane-bound AC (mAC) homologs sharing sequence identities ranging from approximately 50% to 90% (Sunahara et al. 1996) and are complex molecules consisting of two tandem repeats of a hexa-helical transmembrane domain and a cytoplasmic, catalytic CHD. Activity of these NCs requires intramolecular association between the N-terminal CHD (C_1) and C-terminal CHD (C_2) of each cyclase molecule (Dessauer and Gilman 1997; Dessauer et al. 1997; Whisnant et al. 1996; Yan et al. 1996). Studies of mACs have answered some of the basic questions about mechanisms of activity of this family. The function, regulation, and tissue distribution of different mammalian AC homologs have been summarized in several reviews (Cooper et al. 1994; Cooper 2003; Hanoune and Defer 2001; Krupinski and Cali 1998; Smit and Iyengar 1998; Sunahara et al. 1996; Sunahara and Taussig 2002; Tang et al. 1998; Taussig and Zimmermann 1998), and the structure and molecular basis of the catalytic mechanism and regulation of mammalian CHDs have been discussed (Hurley 1998, 1999; Simonds 1999; Tesmer and Sprang 1998). Prokaryotic class III NCs differ considerably in

Table 1 Structures of class III NCs. ATP analogs that have been co-crystallized with CHDs include 5′-(α-thio)-triphosphate (ATPαS), β-L-2′,3′-dideoxy-adenosine-5′-triphosphate (βLddATP) and α,β-methyleneadenosine 5′-triphosphate (AMPCPP). Pyrophosphate (PPi) and P-site inhibitors that have been co-crystallized with the mAC VC1-IIC2 heterodimer include 2′-deoxyadenosine 3′-monophosphate (2d3AMP) and 2′,5′-dideoxy-adenosine-3′-triphosphate (dd3ATP). 2′(3′)-*O*-(*N*-methylanthraniloyl)-guanosine-5′-triphosphate (MANT-GTP) is a nonphysiological inhibitor that binds to the active site of the mAC VC1-IIC2 heterodimer. Crystals for coordinate set 1WC6 contained bicarbonate; see text

PDB ID	Protein	Substrate/ product analogs	Metal A-site	Metal B-site	Regulatory molecules	Reference
Cyclase homology domains						
1AB8	IIC_2	–	–	–	Forskolin	Zhang et al. 1997
1AZS	VC_1-IIC_2	–	–	–	Forskolin, Gs_α	Tesmer et al. 1997
1CJK	VC_1-IIC_2	ATPαS	Mg^{2+}	Mn^{2+}	Forskolin, Gs_α	Tesmer et al. 1999
1CJU	VC_1-IIC_2	βLddATP	Mg^{2+}	Mg^{2+}	Forskolin, Gs_α	Tesmer et al. 1999
1CS4	VC_1-IIC_2	2d3AMP, PP_i	–	Mg^{2+}	Forskolin, Gs_α	Tesmer et al. 2000
1CUL	VC_1-IIC_2	dd3ATP, PP_i	Mg^{2+}	Mg^{2+}	Forskolin, Gs_α	Tesmer et al. 2000
1TL7	VC_1-IIC_2	–	Mn^{2+}	Mn^{2+}	MANTGTP	Mou et al. 2005
1YBT	Rv1900c	–	–	–	–	Sinha et al. 2005
1YBU	Rv1900c	AMPCPP	–	Mn^{2+}	–	Sinha et al. 2005
1Y10	Rv1264	–	–	–	–	Tews et al. 2005
1Y11	Rv1264	–	–	–	–	Tews et al. 2005
1YK9	Rv1625c	–	–	–	–	Ketkar et al. 2006
1FX2	GRESAG 4.1	–	–	–	–	Bieger and Essen 2001
1FX4	GRESAG 4.4	–	–	–	–	Bieger and Essen 2001
1WC0	CyaC	AMPCPP	–	Mg^{2+}	–	Steegborn et al. 2005b
1WC1	CyaC	ATPαS	Mg^{2+}	Mg^{2+}	–	Steegborn et al. 2005b
1WC6	CyaC	ATPαS	Mg^{2+}	Mg^{2+}	–	Steegborn et al. 2005b
1WC5	CyaC	AMPCPP	Displaced Mg^{2+}	Mg^{2+}	–	Steegborn et al. 2005b
2BW7	CyaC	AMPCPP	Displaced Mg^{2+}	Mg^{2+}	Catechol	Steegborn et al. 2005a
GGDEF domains						
1W25	PleD	c-diGMP	–	–	c-diGMP	Chan et al. 2004

length, as their catalytic domains may be linked to a wide variety of other domains responsible for sensing appropriate signals and regulation or interaction with other proteins. In contrast to the mammalian ACs, class III MNCs from lower organisms often have a simpler organization than the mammalian ACs, and typically contain only one CHD.

ACs in the bacteria *Aeromonas hydrophila* and *Prevotella ruminicola* were identified on the basis of their ability to complement *Escherichia coli* cyclase knockout mutants (Cotta et al. 1998; Sismeiro et al. 1998). These ACs do not share detectable sequence similarity with the three classes of NCs described above, or to each other, nor do they appear to contain, in appropriate register, residues that have been shown to be important for function in the class II and class III cyclases. Thus, they have been placed in classes IV and V, respectively (Cotta et al. 1998; Linder and Schultz 2003; Sismeiro et al. 1998). Very recently, the structure of a class IV AC from *Vibrio parahaemolyticus* and *Yersinia pestis* has been solved (Gallagher et al. 2006). Yet another AC—CyaC from *Rhizobium etli*, also identified by its ability to restore AC activity in *E. coli* cyclase knockout mutants—has been proposed to constitute a sixth class of cyclases (Tellez-Sosa et al. 2002). However, several key residues required for binding and catalysis by the class III NCs are present at similar relative positions in this protein, although it shares less than 15% sequence identity with other class III NCs that have been studied (Fig. 1). Thus, in the absence of further structural, mutational, or biochemical evidence, it is likely that this enzyme and other close homologs may simply constitute yet another subfamily of class III NCs.

Structural homology of class III NCs

The first NC structures determined were those of class III CHDs belonging to the mammalian ACs. The core fold of the CHDs was determined from the structure of the mammalian IIC_2 domain (Zhang et al. 1997) and verified by subsequent structures of CHDs (Table 1; Fig. 2a). The core fold of the GGDEF domains was determined more recently from the *Caulobacter crescentus* PleD structure (Fig. 2b; Chan et al. 2004). As has been previously discussed, numerous proteins, including the palm domain of type I DNA polymerases and the catalytic domains of class III NCs share a ferredoxin-like βαββαβ structural motif, such that the αββ connection between the two βαβ repeat units is left-handed (Fig. 2c; Artymiuk et al. 1997; Murzin 1998; Pei and Grishin 2001). This motif corresponds to β1-α2-β2-β3-α3-β4 of the class III NC core fold, in which the strands form a part of the central β-sheet (β2-β3-β1-β4). In all class III NCs and the palm domain of type I DNA polymerases, a short helix, α1, is inserted between β1 and α2, and the α2 helix is long. Thus, these two protein families have a common core βααββαβ structural motif (Fig. 2a–c).

Two aspartic acid residues that are invariant among CHDs have been shown to be critical for catalysis. These residues are located on the core βααββαβ structural motif; the first is the penultimate residue of β1, while the second is located at the C-terminus of the β2-β3 loop (Figs. 1 and 3a). These two aspartates are also very highly conserved in the GGDEF domains, although in many homologs the β2-β3 loop aspartate, which corresponds to the central residue of the GGDEF motif, is substituted by a glutamate (Figs. 1 and 3b). The core βααββαβ structural motif of the type I DNA polymerases also bears two invariant aspartates critical for catalysis at topologically equivalent positions (Fig. 3c). The conservation of the βααββαβ structural motif (Fig. 2), combined with the preservation of catalytically important residues (Fig. 3) in class III NCs and type I DNA polymerases, suggests that these two enzyme families share a common ancestor.

The core βααββαβ structural motif of class III NCs is further extended by a helix (α4) and another strand (β5) (Figs. 1 and 2a–b). Thus, as was previously predicted, the CHDs and GGDEF domains have a common order of core secondary structure elements consisting of β1-α1-α2-β2-β3-α3-β4-α4-β5, that folds into a compact α/β sandwich (Pei and Grishin

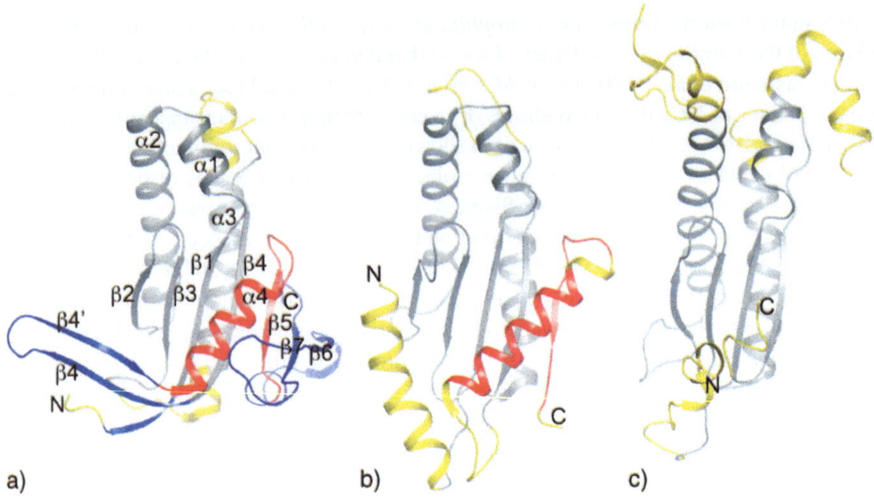

Fig. 2a–c The βααββαβαβ structural motif conserved in class III NCs and DNA polymerases. Secondary structure elements are colored and labeled as in Fig. 1. This and all molecular figures were made using the program PYMOL (http://pymol.sourceforge.net/). **a** CHD fold with secondary structural elements labeled. **b** GGDEF domain fold. **c** Palm domain of DNA polymerase

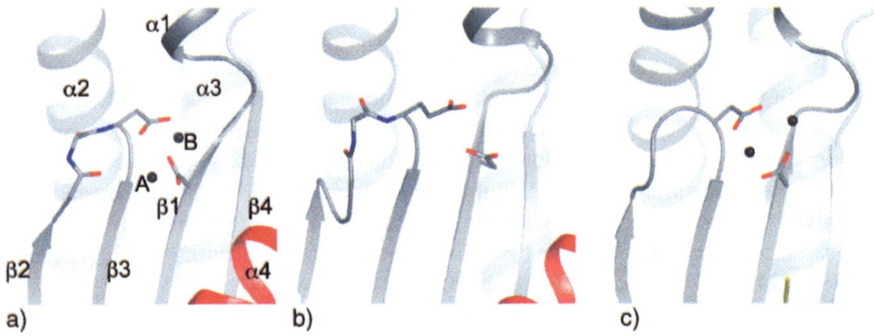

Fig. 3a–c Structural equivalence residues highly conserved in class III NCs and DNA polymerases. Secondary structure elements are colored and labeled as in Fig. 1. Conserved residues—including the two acidic residues conserved in both these enzyme families, and the β2-β3 loop glycine conserved in class III NCs—are shown in molecular detail. Metal ions are indicated by *black spheres*. **a** CHD. **b** GGDEF domain. **c** Palm domain of type I DNA polymerase

2001). This catalytic core fold is conserved among all class III NCs and consists of a central, five-stranded, antiparallel β-sheet of strand order β2-β3-β1-β4-β5, with three helices, α1–α3, on the back face of this sheet and another helix, α4, on the front face (Figs. 1 and 2a–b). In addition to the two invariant acidic residues on β1 and the β2-β3 loop, two conserved glycines are located at structurally equivalent positions in all class III NCs (Figs. 1 and 3a–b). The first glycine precedes the conserved acidic residue on the β2-β3 loop. Thus, the β2-β3 loop of all class III NCs bears a conserved Gly-Asp pair, which in DGCs corresponds to the second residue of the GGDEF motif. The Φ,Ψ dihedral angles of this glycine map to unfavorable regions of the Ramachandran plot in all available class III NC structures, indicating that a glycine at this position may be important for maintaining a specific conformation of the β2-β3 loop. The second glycine conserved in class III NCs maps to the middle of β4 and is sometimes substituted by other small residues such as alanine. This glycine is conserved due to steric constraints arising from the packing of α4 against β4 (Pei and Grishin 2001). This β4 glycine is not conserved in the palm domains of type I DNA polymerases, as the core βααββαβ structural motif of these domains is not extended by the α4-β5 extension present in class III NCs.

Distinct insertions into the conserved class III NC core fold further distinguish the CHDs from the GGDEF domains. The GGDEF domain of *C. crescentus* PleD is modified by a subdomain composed of a β-hairpin (β0-β0′) and a helix (α0) that precedes the class III NC core fold as well as by another β-hairpin (β3′-β3″) inserted between β3 and α3 of the class III NC core (Fig. 2b). The function and conservation of these insertions among the GGDEF domains remains to be fully investigated. However, sequence alignments suggest that the β3′-β3″ is absent among archaeal and some bacterial GGDEF domains and may be a distinguishing feature between subgroups of GGDEF domains (Pei and Grishin 2001).

The class III NC core fold in all CHDs is modified by two insertions, which constitute the most variable features of CHDs (Figs. 1, 2a, and 4). The first of these insertions, named the dimerization arm, is introduced between β4 and α4, and consists of a β-ribbon comprising C-terminal residues of the extended β4 strand and an antiparallel β-strand (β4′). The dimerization arm is bounded by two almost-invariant glycines (Fig. 1). The first of these glycines corresponds to the last β4 residue within the class III NC core, while the second, which is occasionally substituted by a serine, precedes α4 of the class III NC core at the end of the dimerization arm (Fig. 1). The Φ,Ψ dihedral angles for both these glycines map to unfavorable regions of the Ramachandran plot in all available CHD structures, indicating that glycines at these position facilitate the bending of β4 and the tight turn between the dimerization arm and α4. The length of this dimerization arm defined by these two invariant glycines ranges from 10 to 19 residues and constitutes an important distinguishing feature among the four CHD subclasses designated by Linder and Schultz (Figs. 1 and 4; Linder and Schultz 2003). In some homologs, the N-terminal residues of β1 are also hydrogen-bonded to the β-ribbon of the dimerization arm, forming a three-stranded β-sheet of strand order β1-β4-β4′ (Fig. 2a). The second CHD-specific insertion follows β5 of the class III NC core fold and consists of a helix, α5, followed by an antiparallel β-hairpin, β6-β7 (called β7-β8 in structures of CHDs) that extends the central β-sheet by two strands. Compared to other CHD structures, β5 is shorter in mammalian VC_1 domains, and the β6-β7 hairpin is replaced by an Ω loop consisting of two short α-helices (Tesmer et al. 1997). A glycine, whose Φ,Ψ dihedral angles also map to generously allowed regions of the Ramachandran plot, is almost invariant on the β6-β7 loop of all CHDs in which the following residue is a basic residue that hydrogen-bonds the polyphosphate group of substrate or product, suggesting that this glycine assists in positioning the basic residue in these CHDs (Fig. 1).

According to the Linder and Schultz classification, the mammalian VC_1 and IIC_2 domains and the Rv1625c CHD belong to class IIIa, *S. platensis* CyaC belongs to class IIIb, *M. tuberculosis* Rv1900c and Rv1264 belong to class IIIc, and the trypanosomal GRESAG ACs belong to class IIId (Linder and Schultz 2003). Thus, structures of CHDs from nearly the full spectrum of class III NCs have now been determined (Fig. 4 and Table 1). All of these CHDs have the same CHD core fold, but have diverged with respect to insertions such as the length of the β4-β4′ ribbon (dimerization arm), the length and structure of the α5-β6-β7 subdomain, and the presence of variable insertions such as the δ-subdomain of the trypanosomal GRESAG proteins (Fig. 4).

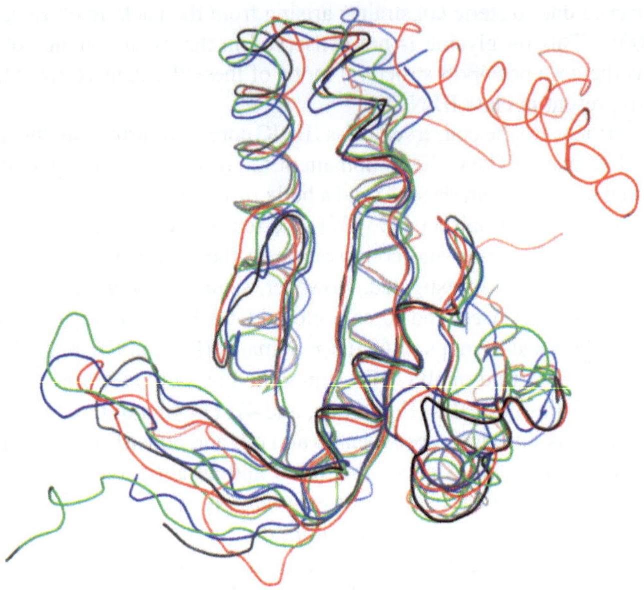

Fig. 4 CHDs from subclasses IIIa–d. Structural elements constituting the core class III NC fold are much better conserved than the CHD-specific features. PDB IDs of the superimposed CHDs from subclasses a–d are CanineVC_1 (1CJK_A) in *green*, *S. platensis* CyaC (1WC6) in *blue*, *M. tuberculosis* Rv1900c (1YBU) in *black*, and Trypanosomal GRESAG 4.1 (1FX2) in *red*, respectively

Quaternary structure of class III NCs

Structures of the catalytically competent VC_1-IIC_2 heterodimers in complex with substrate analogs or inhibitors (Tesmer et al. 1997) conclusively demonstrated that CHDs are active only as dimers, as had been suggested on the basis of solution studies of these domains (Dessauer and Gilman 1997; Dessauer et al. 1997; Scholich et al. 1997a; Tang and Gilman 1995; Whisnant et al. 1996; Yan et al. 1996). The GGDEF domains have not been studied as extensively, but the recent structure of *C. crescentus* PleD, in complex with the c-diGMP, suggests that dimerization is a prerequisite for catalysis by these enzymes as well (Chan et al. 2004).

The mammalian ACs were the first NCs to be extensively studied. Structures of catalytically competent VC_1-IIC_2 heterodimers demonstrated that CHDs form intramolecular, isologous dimers (Tesmer et al. 1997). As the two CHDs are homologous, but have non-identical sequences and associated structural variations, these intramolecular heterodimers are pseudo-symmetrical. Class III NCs from lower organisms typically contain only one CHD and are active as intermolecular homodimers (Guo et al. 2001; Linder and Schultz 2003). Structures of three homodimeric NCs, two from *M. tuberculosis* and one from the cyanobacterium *S. platensis*, indicate that these homodimeric CHDs may form either pseudo-symmetrical or perfectly symmetrical, isologous homodimers, depending on structural differences between the chemically identical CHDs (Sinha et al. 2005; Steegborn et al. 2005b; Tews et al. 2005). Thus, packing interactions between dyad-related CHDs may be symmetrical or pseudo-symmetrical.

Dimerization contacts between CHDs are extensive, burying 1,500–3,500 Å of the total accessible surface area at the dimer interface and involving both hydrophobic and electrostatic interactions. One face of the dimer, termed the ventral face (the side of the dimer from which substrates can enter the catalytic site), bears a shallow trough that runs along the dimer interface perpendicular to the dyad axis (Fig. 5a). The β2-β3 hairpins from each subunit abut each other at the center of the dimer interface, dividing this trough into two nucleotide-binding sites related by the dyad axis. While the two β2-β3 hairpins always form the inner edge of each site, the outer edge is usually bordered by the dimerization arm of one CHD packed against α2, the α1-α2 turn and in some structures, β2 of the other CHD. Additionally, the β4′-α4 turn of one subunit contacts the α1-α2 turn of the other subunit (Fig. 5a). In some CHDs dimerization may be significantly perturbed by mutations on the dimerization arm or the β2-β3 hairpin (Ketkar et al. 2004, 2006; Shenoy et al. 2003). There are significant variations in these packing interactions, not only between dimers of different CHDs, but also between the dyad-related elements of pseudo-symmetrical dimers. Notably, compared to the class IIIa and class IIIb CHDs, the dimerization arms of the class IIIc CHDs are shorter (Fig. 1; Linder and Schultz 2003) and make fewer inter-domain contacts, accounting for a decrease of approximately 1,500 Å of surface area buried at the dimer interface of these CHDs (Sinha et al. 2005; Tews et al. 2005).

Each of the two dyad-related sites at the CHD dimer interface constitutes a potential active site (Fig. 5a). Residues from each domain line both potential active sites. The β2-β3 hairpin of each CHD, which participates in both sites, divides each CHD into two structural modules. The first module comprises the β1-α1-α2-β2-β3-α3-β4 core structural motif that is conserved among class III NCs and type I DNA polymerases and provides most of the determinants for binding divalent metal cations and polyphosphate groups, as well as elements essential for chemical reactions involving a 3′-5′-phosphodiester bond. The second module is composed of the α4-β5 core fold extension present in all the class III NCs, as well as the two CHD-specific insertions, the β4-β4′ dimerization arm and the α5-β6-β7 extension (Fig. 2a) and provides most of the elements for binding the nucleotide base and ribose. Within each CHD, the first module contributes residues to one binding site, while residues from the second module participate in the dyad-related binding site.

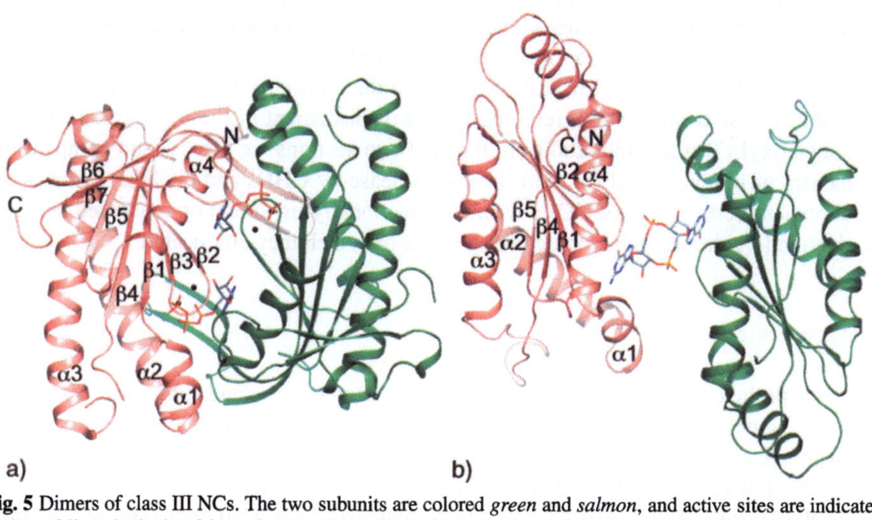

Fig. 5 Dimers of class III NCs. The two subunits are colored *green* and *salmon*, and active sites are indicated by bound ligands depicted in molecular detail. Secondary structure elements and termini of the polypeptide chain of one subunit are labeled as in Fig. 1. **a** CHD domain catalytic dimer. **b** Proposed GGDEF domain catalytic dimer. Note the different dyad axes relating the subunits of each of these dimers

Oligomerization of GGDEF domains in their active state in solution has not been clearly established. However, the *C. crescentus* PleD structure in complex with the product c-diGMP indicates that a single GGDEF domain provides binding determinants for only one GTP, or half of a c-diGMP molecule, suggesting that dimerization is essential for activity (Chan et al. 2004). Indeed, the binding determinants for the other half of the c-diGMP molecule are provided by a second PleD molecule, related to the first by the dyad axis of the c-diGMP (Fig. 5b). However, it is not certain whether the dimers observed in crystals correspond to a biologically relevant arrangement of subunits as, unlike most physiological dimers, the two subunits do not make any direct protein–protein interaction and are linked only by the shared c-diGMP molecule (Fig. 5b). The paucity of interactions between subunits is also in contrast to the extensive dimerization contacts seen between subunits in the CHD dimers. Although, like the CHD homodimers, the active site of GGDEF domains is also located at the dimer interface, unlike the CHD homodimers, each GGDEF domain contributes all the elements to bind one GTP molecule (Fig. 5). Further, unlike the CHD homodimers, which have two chemically equivalent active sites, the GTP-binding sites of two GGDEF domains combine to form a single active site (Fig. 5). The proposed two-fold symmetry axis that relates subunits in a catalytically competent GGDEF domain dimer is different from that which relates subunits in a CHD dimer.

Active sites and catalytic mechanism

NCs catalyze the synthesis of 3′,5′ cNMPs from their respective NTPs. Most NCs are specific for either ATP or GTP. The cyclization reaction has been observed to proceed with an inversion of stereochemistry in the class III mammalian ACs, suggesting that it involves the in-line nucleophilic attack of the ATP 3′-hydroxyl on the 5′ α-phosphate, leading to a pentavalent-phosphate transition state intermediate. and subsequently to the products cAMP and pyrophosphate (Eckstein et al. 1981). It is likely that this chemical reaction mechanism is common to all class III NCs. Thus, all class III NC active sites are constrained by the common objectives of binding NTP, distinguishing between ATP and GTP, increasing the nucleophilicity of the 3′-hydroxyl, as well as a mechanism that facilitates the release of products. Further, as the cyclization reaction catalyzed by the MNCs is intramolecular, the active sites of these enzymes are expected to have the ability to stabilize the strained transition state ribose conformation, required for the displacement of the 5′ pyrophosphate by the 3′-hydroxyl (Tesmer et al. 2000). In the next two sections we discuss and compare the mechanism of binding, catalysis, and substrate release by CHDs and GGDEF domains. In all subsequent descriptions of NC active sites, functionally important or conserved residues are identified by the secondary structure elements on which they are located.

CHD active sites

Structures of the catalytically competent VC_1-IIC_2 heterodimers, and subsequently of CHD homodimers, demonstrated that, in both hetero- and homodimers the active site is located in a deep groove at the interface of the CHDs (Sinha et al. 2005; Steegborn et al. 2005b; Tesmer et al. 1997; Tews et al. 2005). Structures of the mammalian VC_1-IIC_2 CHD heterodimers as well as the *M. tuberculosis* Rv1900c and the *S. platensis* CyaC CHD homodimers, in complex with various ligands (Table 1) have enabled investigators to elucidate the general mode by which CHDs bind the substrate ATP (Fig. 6a). NC active sites require elements to bind

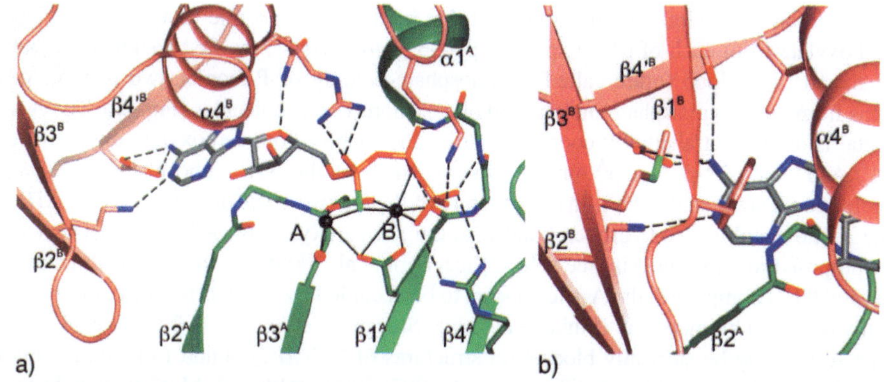

Fig. 6a, b Active site of class III MNCs. The two CHDs are colored as in Fig. 5 and secondary structures labeled as in Fig. 1. Secondary structures from the module comprising the core βααββαβ structural motif of one CHD are denoted by a *superscripted A* (^{A}), while those from the second module comprised of class III NC-specific or CHD-specific elements from the other CHD are denoted by a *superscripted B* (^{B}). Molecular details of bound ligands and of protein residues that play a role in substrate binding and catalysis are shown. *Solid, black lines* represent the coordination sphere of the bound metals while *dashed, black lines* indicate hydrogen bonds. **a** ATP analog ATPαS-bound active site. Protein elements that obstruct this view of the active site are not shown. **b** Purine-binding sub-site

the negatively charged polyphosphate group and stabilize the ribose and purine moieties of the NTP. In CHDs the former role is primarily performed by electrostatic interactions with basic amino acids and coordination by divalent metal cations, which in turn are coordinated by carboxylate and carbonyl groups from the protein. In contrast, the mechanism of binding the ribose and purine appear to be less specific and therefore less well conserved, chiefly involving hydrophobic packing and stacking interactions, although polar interactions play an important role in purine recognition. As described above, residues from each CHD contribute to both dyad-related binding sites (Sinha et al. 2005; Steegborn et al. 2005b; Tesmer et al. 1997). Thus, each dyad-related site is formed by residues from the β2-β3 loop of both CHDs, residues from the first module comprising the core βααββαβ structural motif of one CHD denoted by a superscripted A (^{A}), and residues from the second module made up of class III NC-specific or CHD-specific elements from the other CHD, denoted by a superscripted B (^{B}) (Fig. 6a).

Coordination of metal in CHD active sites

Like most other enzymes that catalyze reactions involving phosphodiester bonds, such as RNA and DNA polymerases and ribozymes, CHDs require divalent metal cations for catalysis (Beese and Steitz 1991; Doublié and Ellenberger 1998; Steitz 1993, 1999; Steitz et al. 1994; Steitz and Steitz 1993; Tesmer et al. 1997; Zimmermann et al. 1998). Based on evidence from mutagenesis (Zimmermann et al. 1998), and crystal structures of the VC_1-IIC_2 heterodimer in complex with ATP analogs R_p-ATPαS or βLddATP (Tesmer et al. 1999), the active site binds two divalent metal cations (Fig. 6a). These metal cations occupy sites referred to as "A" and "B," in analogy to corresponding sites in DNA and RNA polymerases (Doublié and Ellenberger 1998; Tesmer et al. 1999). The B-site is typically occupied by Mg^{2+} or Mn^{2+} in structures of CHDs to which NTPs (or pyrophosphate) are bound, indicating that binding of metal at this site is a prerequisite to binding the β and γ phosphates of the substrate ATP, or the by-product of ATP cyclization, pyrophosphate (Mou et al. 2005; Sinha et al. 2005; Steegborn et al. 2005b; Tesmer et al. 1999). The B-site metal cation is typically bound in an octahedral coordination shell that includes at least one (but sometimes both) carboxylate from each of $\beta1^A$ Asp^A and $\beta2^A$-$\beta3^A$ loop Asp^A, the first backbone carbonyl of the $\beta1^A$-$\alpha1^A$ loop (often called the phosphate-binding or P-$loop^A$), two or three oxygens from the polyphosphate moiety, and one or more water molecules (Fig. 6a). The A-site metal is usually tetrahedrally coordinated by either one or both carboxylate oxygens from $\beta1^A$ Asp^A and $\beta2^A$-$\beta3^A$ Asp^A, one or two oxygens from the ATP α-phosphate, and a water molecule (Tesmer et al. 1999). In the structure of *S. platensis* CyaC in complex with ATPαS, the A-site metal is also hydrogen-bonded to the 3′-hydroxyl of R_p-ATPαS, consistent with its proposed role in catalysis (see below; Steegborn et al. 2005b). In the crystal structures of several CHD complexes, the A-site appears to be occupied by a water molecule, or only partially occupied by metal ion (Sinha et al. 2005; Steegborn et al. 2005b; Tesmer et al. 1999). The A-site may be sterically blocked in structures of VC_1:IIC_2 bound to P-site inhibitors with 3′ phosphate substituents (Tesmer et al. 1997, 2000). Although Mg^{2+} is thought to be the physiologically relevant cofactor for most CHDs, bonds constituting the coordination shell of each metal site are longer than those typical for Mg^{2+}. This may explain the enhanced activity of most CHDs in the presence of larger divalent metal cations like Mn^{2+}. Further, structural evidence indicates that the A-site preferentially binds and is inhibited by Zn^{2+}, perhaps because a tetrahedral coordination geometry is inherently favored at this site, while the B-site preferentially binds metal cations that favor an octahedral coordination geometry such as Mn^{2+}, Ca^{2+}, Sr^{2+}, and Eu^{2+} (Steegborn et al. 2005b; Tesmer et al. 1999). The two aspartates responsible for coordinating metal, $\beta1^A$ Asp^A and $\beta2^A$-$\beta3^A$ loop Asp^A, are invariant among most CHDs and constitute the best-conserved features of CHD active sites. This emphasizes the importance of metal cations, not only in binding ATP, but also for catalysis. The importance of these aspartates, and implicitly the metals they coordinate, is underscored by evidence that mutations of these residues dramatically reduce activity (Liu et al. 1997; Sinha et al. 2005; Tang et al. 1995; Tesmer et al. 1999). The two metal sites are separated by 3.9 Å in most enzymes that utilize two divalent metal cations to catalyze chemical reactions involving phosphodiester bonds (Beese and Steitz 1991; Doublié and Ellenberger 1998; Steitz 1993, 1999; Steitz and Steitz 1993). Consistent with this, the two sites are separated by 3.5–4.5 Å in most structures of the VC_1-IIC_2 heterodimer in complex with various ligands, as well as that of *M. tuberculosis* Rv1900c CHDs in complex with AMPCPP, where the A-site is occupied by water (Sinha et al. 2005; Tesmer et al. 1999, 2000). However, considerable variation in the placement and ligation of Mg^{2+} is observed in the structures of CHDs in complex with certain nonphysiological inhibitors, such as that of VC_1:IIC_2 bound to MANT-GTP (Mou et al. 2005), and of *S. platensis* CyaC in complex with AMPCPP (Steegborn et al. 2005b).

Phosphate binding in CHD active sites

Binding of phosphate groups in CHD active sites is quite variable and a variety of ligands, including nucleotide 5′ monophosphates, nucleotide 3′ monophosphates, nucleotide 3′ triphosphates, and pyrophosphate, can be accommodated (Table 1). Polyphosphate moieties of substrate analogs and the product pyrophosphate are bound to essentially the same subsite of the CHD active site. This subsite is located at the C-terminal edge of the central sheet ($\beta2^A$-$\beta3^A$-$\beta1^A$-$\beta4^A$) of the conserved βααββαβ structural motif (Fig. 6a). The conformation of the polyphosphate groups bound in this subsite, as well as details of interactions with protein groups, varies substantially among different ligand complexes, although some general features are preserved. In general, the polyphosphate group binds such that the α- and β-phosphates are extended over the $\beta2^A$-$\beta3^A$ loop and the γ-phosphate is stabilized over a positively charged pocket formed by the three-residue P-loop^A ($\beta 1^A$-$\alpha 1^A$ loop) and the helix dipole of $\alpha 1^A$ (Fig. 6a). The phosphate groups are stabilized via coordination by metal, as well as by hydrogen bonds to various positively charged protein groups such as the amides of the P-loop^A and conserved basic side-chains.

As mentioned in "Coordination of metal in CHD active sites," binding of metal at the B-site is a prerequisite for binding polyphosphate. The B-site metal usually coordinates and stabilizes both β- and γ-phosphates and often the α-phosphate as well. In contrast, the A-site metal most often coordinates the α-phosphate, and only occasionally either the ribose-α phosphate bridging oxygen or the β-phosphate. Backbone groups of the three-residue P-loop^A, are involved in binding the polyphosphate. The characteristic conformation of this loop is stabilized by hydrophobic packing interactions of the last β1 residue, which is usually a conserved isoleucine (Fig. 1). The main chain carbonyl of the last β1 residue and main chain amides of the first and second residues of the P-loop hydrogen-bond, either one or both, the β- and γ-phosphates. Three highly conserved basic residues—$\beta 4^A$ Arg^A, $\alpha 4^B$ Arg^B, and $\beta 6^B$-$\beta 7^B$ Lys^B—also play key roles in stabilizing the polyphosphate (Figs. 1 and 6a). The $\beta 4^A$ Arg^A appears poised to form bidentate salt bridges with the γ-phosphate, but is located just beyond optimal hydrogen-bonding distance in most structures of CHD dimers in complex with ligands that contain polyphosphate. The $\beta 6^B$-$\beta 7^B$ Lys^B usually ion pairs with the γ-phosphate, or sometimes with either the α- or β-phosphate. Mutation of either of these residues usually increases K_m, consistent with their role in binding substrate (Dessauer et al. 1997; Sinha et al. 2005; Tang et al. 1995). The $\alpha 4^B$ Arg^B often hydrogen-bonds the α-phosphate or occasionally the α-β bridging oxygen and has been proposed to stabilize the pentavalent α-phosphate transition state intermediate (Sinha et al. 2005; Tesmer et al. 1999). Consistent with this role, mutating this residue usually shows the rate of reaction, but does not impact affinity for substrate (Sinha et al. 2005; Tang et al. 1995; Yan et al. 1997b).

In structures of *S. platensis* CyaC in complex with substrate analogs, a conserved asparagine from $\alpha 4^B$ hydrogen-bonds the oxygen bridging either the ribose and α-phosphate, or the α- and β-phosphates (Steegborn et al. 2005b). However, this may be a unique feature of *S. platensis* CyaC, as equivalents of this $\alpha 4^B$ Asn^B do not make equivalent contacts in the mammalian VC_1-IIC_2 heterodimers or Rv1900c CHD homodimers. Finally, in addition to these conserved contacts, the polyphosphate group in different protein complexes has been observed to form additional interactions, which include water-mediated hydrogen bonds to different protein groups, as well as direct hydrogen bonds to nonconserved residues.

Binding of ribose in CHD active sites

The ribose ring is bound in the space between the $\beta 2^A$-$\beta 3^A$ loop and $\alpha 4^B$ (Fig. 6a). Like the pyrophosphate moiety, the conformation of the ribose ring varies, and both 2′- and 3′-endo conformations have been observed. Further, in the structures of VC_1-IIC_2 in complex with βLddATP, the ribose ring is inverted relative to the orientation seen in complexes of CHDs with ligands such as R_p-ATPαS or AMPCPP, as well as P-site inhibitors (Tesmer et al. 1999). However, in contrast to the polyphosphate, the ribose ring makes few direct contacts with the protein, and these are conserved poorly among CHDs (Figs. 1 and 6a). The 2′ and 3′-hydroxyls usually do not directly contact any protein groups, but each may be involved in one to three, nonconserved, water-mediated bonds. However, in structures of *S. platensis* CyaC in complex with two Mg^{2+} and AMPCPP or R_p-ATPαS, the 3′-hydroxyl is hydrogen-bonded to the $\alpha 4^B$ Arg^B or the A-site metal, respectively (Steegborn et al. 2005b). While the mechanistic relevance of the former is unclear, the latter is clearly pro-catalytic as discussed later in the context of the catalytic mechanism. The ribose ring oxygen is hydrogen-bonded to amide of the $\alpha 4^B$ Asn^B in structures of the mammalian VC_1-IIC_2 heterodimers in complex with R_p-ATPαS, and is implicated in orienting the substrate or transition state (or both) for catalysis (Tesmer et al. 1999). This conclusion is supported by mutations of the mammalian enzymes (Yan et al. 1997b). In *M. tuberculosis* Rv1900c CHDs, $\alpha 4^B$ Asn^B is substituted by a histidine, which does not participate in direct interactions with ATP. Mutational evidence further confirms that this residue is not required for catalysis by Rv1900c CHDs (Sinha et al. 2005). Structures of *S. platensis* CyaC indicate that in these enzymes, the $\alpha 4^B$ Asn^B is present in a location topologically equivalent to that in the mammalian VC_1-IIC_2 heterodimers, but is not located within hydrogen-bonding distance of the ribose oxygen of the bound NTP. It is possible that further closure of the active site is required to allow formation of this hydrogen bond or, as in Rv1900c, this residue does not play a role in catalysis. As equivalents of $\alpha 4^B$ Asn^B are variously substituted or missing in many CHDs identified from genome sequences, it is likely that either the specific elements for stabilizing the ribose ring and orienting substrate are not general prerequisites of the CHD active site or that these elements vary greatly among the CHDs.

Base recognition by CHD active sites

Selection of correct nucleotide substrates by NCs is essential to the fidelity of cyclic nucleotide-mediated signal transduction. In contrast to the ribose and phosphate, the purine ring is inflexible and it appears to form identical interactions in different nucleotide complexes with the same enzyme. However, there is substantial variation in the mechanism of purine recognition used by different CHDs. In all CHDs, the purine base binds in a hydrophobic pocket formed by conserved residues from the $\beta 2^B$-$\beta 3^B$ hairpin, $\beta 1^B$, $\beta 2^B$, $\alpha 4^B$, and the dimerization arm^B(Figs. 1 and 6b). The hydrophobic packing interactions include edge-to-face stacking with a conserved aromatic residue from $\beta 1^A$ that precedes the invariant $\beta 1^A$ Asp^A by two residues (Figs. 1 and 6b). In addition to these interactions, the purine ring is stabilized by stacking against either one of the two peptide planes that involve the $\beta 2^A$-$\beta 3^A$ loop glycine that is invariant in all class III NCs (Figs. 1 and 6b). There are few polar interactions with the purine ring, and these serve chiefly to discriminate between ATP and GTP (Figs. 1 and 6b). Due to the nonspecific nature and variability of many of these interactions, it has been difficult to completely understand the mechanism of base recognition by CHDs. Variations in the modes by which CHDs accomplish this function have been summarized in a recent review (Linder 2005). In this section, we outline general features but describe only results of structure-based studies in some detail.

Typically, ACs are highly specific for ATP with no detectable activity with GTP (Coudart-Cavalli et al. 1997; Guo et al. 2001; Kasahara et al. 2001; Linder et al. 2002, 2004; Shenoy et al. 2005; Sunahara et al. 1998; Weber et al. 2004), although Cya1, an AC from *Rhizobium meliloti* has detectable GC activity (Beuve et al. 1993). In comparison, various GCs have been shown to have significant AC activity also (Beuve 1999; Linder et al. 1999, 2000; Sunahara et al. 1998; Tucker et al. 1998). Two residues conserved among AC CHDs—the $\beta 2^B$-$\beta 3^B$ hairpin Lys^B, which precedes the invariant β2-β3 loop Gly-Asp by two residues, and the dimerization arm^B Asp^B—were identified as key determinants of base specificity from structures of the mammalian VC_1-IIC_2 heterodimers (Figs. 1 and 6b; Tesmer et al. 1997). These two residues hydrogen-bond the N1 and N6 of the ATP adenine and preclude binding of GTP. Recent structures demonstrate that in *S. platensis* CyaC, ATP is similarly selected by hydrogen bonds between the topologically equivalent residues, a $\beta 2^B$-$\beta 3^B$ hairpin Lys^B and a dimerization arm^B Thr^B and the N1 and N6 of the ATP adenine (Steegborn et al. 2005b). In the mammalian ACs, a third residue—a dimerization arm^A Gln^A—that in VC$_1$ precedes the equivalent of IIC$_2$ dimerization arm^B Asp^B by two residues (Fig. 1) and is within the van der Waals contact radius of the β2^B-β3^B hairpin Lys^B, has been proposed to play a role in substrate selection (Sunahara et al. 1998). Structures of GCs are not yet available, but the sequence equivalents of β2^B-β3^B hairpin Lys^B and dimerization arm^B Asp^B of the mammalian ACs are conserved among GCs (Fig. 1) as a glutamate and cysteine, respectively, and are proposed to hydrogen-bond the N2 and O6 of guanine (Fig. 1; Beuve 1999; Sunahara et al. 1998; Tucker et al. 1998). A GC arginine, which is the equivalent of the VC$_1$ dimerization arm^A Gln^A, is proposed to orient the conserved glutamate by salt bridges as well as packing interactions. However, mutational analyses suggest that these residues may not be the sole determinants of base specificity in these enzymes. Mutating the mammalian IIC$_2$ β2^B-β3^B hairpin Lys^B, the dimerization arm^B Asp^B, and the dimerization arm^A Gln^A to glutamate, cysteine, and arginine, respectively, does not switch specificity of the VC$_1$-IIC$_2$ heterodimer, but rather converts it into a nonspecific NC with reduced activity (Sunahara et al. 1998). Mutations of the β2^B-β3^B hairpin Lys^B and the dimerization arm^B Asp^B in ACs from lower organisms often abolish all enzyme activity (Kasahara et al. 2001; Ketkar et al. 2006; Shenoy et al. 2003, 2005; Sunahara et al. 1998). In contrast to ACs, mutation of the glutamate, cysteine, and arginine found in GCs to a lysine, aspartate, and glutamine, respectively, converted GCs to a stringently ATP-specific AC (Beuve 1999; Linder et al. 2000; Sunahara et al. 1998; Tucker et al. 1998). In fact, the single conversion of cysteine to an aspartate was adequate for switching specificity, suggesting that the dimerization arm^B Asp^B is the most important determinant for substrate specificity while the other residues play supporting roles.

The recently characterized mycobacterial nucleotidyl cyclase Rv1900c can use both ATP and GTP as substrate, with a 14-fold preference for ATP (Sinha et al. 2005). Apart from Rv1900c, only one other prokaryotic enzyme has been shown to have GC activity (Ochoa de Alda et al. 2000). Strikingly, in Rv1900c the mammalian IIC$_2$ β2^B-β3^B hairpin Lys^B that hydrogen-bonds the adenine N1 is replaced by an asparagine and may account for its reduced specificity, with the dimerization arm^B Asp^B establishing the preference for ATP. However, structures of Rv1900c in complex with AMPCPP indicated that neither the β2^B-β3^B loop Asn^B, nor the dimerization arm^B Asp^B hydrogen-bond the adenine N1 or N6, and therefore are probably not the main determinants for the preferential use of ATP as substrate. Instead of hydrogen-bonding to the adenine, the β2^B-β3^B loop Asn appears to sterically enforce binding of the purine ring in a half-of-sites binding mechanism discussed in "Evolution of Class III NC active sites." Correspondingly, the mutation of β2^B-β3^B loop Asn^B does not significantly impact substrate specificity or enzyme activity, although inexplicably the dimerization arm^B Asp^B mutants have higher affinity for both ATP and GTP. The preference for ATP in Rv1900c has been attributed to nonlocalized general determinants such as electrostatic gradients from helix dipoles or peptide planes, shape complementarity, and size of the active site (Sinha et al. 2005). Additionally, interactions with backbone groups, such as the hydrogen bond between the adenine N6 and the carbonyl of the residue following the dimerization arm^B Asp^B observed in structures of VC$_1$-IIC$_2$ heterodimers may further select for ATP, and against GTP (Tesmer et al. 1997, 1999). It is likely that such general determinants play a role in purine specificity in all CHDs. In ACs, such general determinants probably dictate the preference for ATP, while the β2^B-β3^B hairpin Lys^B–dimerization arm^B Asp^B couple enforce the strict specificity for ATP. In contrast, general determinants in the active site of GCs appear to allow binding of both ATP and GTP, with the preference for GTP dictated chiefly by the glutamate-cysteine couple.

Catalysis by CHDs

Although the structures of CHDs in complex with various ligands tell us much about binding the substrate ATP, none of these structures completely mimics the binding of the putative transition state. The product cAMP, which is rapidly released, may be the best analog. The transition state intermediate is expected to have a 3′-endo ribose conformation that positions the 3′-hydroxyl for in-line displacement of the pyrophosphate, with a pentavalent α-phosphate, whose negative charge is stabilized by the $\beta4^B$ Arg^B. As in many DNA polymerases, ribozymes, and phosphatases that employ a mechanism of two-metal ion catalysis, the A-site metal is expected to play an indispensable role in catalysis by CHDs. As the 3′-hydroxyl is a weak nucleophile, it has been proposed that metal bound at the A-site increases the nucleophilicity of the ATP 3′-hydroxyl (Tesmer and Sprang 1998). Indeed, consistent with the predicted role of the A-site metal as a Lewis acid, in structures of *S. platensis* CyaC in complex with two Mg^{2+} and R_p-ATPαS, the ribose 3′-hydroxyl is hydrogen-bonded to the A-site metal (Steegborn et al. 2005b). However, it is unlikely that this structure completely mimics the transition state, as the 3′-hydroxyl is not positioned for an in-line attack on the α-phosphate. In contrast to the nearly invariant orientation of the purine ring in substrate analogs bound to CHDs, the conformational flexibility of the ribose and pyrophosphate groups—both between different enzymes as well as between different structures of the same enzyme—may be indicative of the flexibility required to attain the strained transition state intermediate. Consequently, the variability captured in different crystal structures may represent an inherent feature of CHD active sites.

Product binding and release by CHDs

There are no structures of CHDs in complex with the products cAMP and pyrophosphate, perhaps because the same mechanism that enables release of products in the normal course of catalysis prevents the product-bound state from being captured in a crystal structure. In the mammalian ACs, cAMP appears to be released first, and the release of the pyrophosphate appears to be the rate-limiting step (Dessauer and Gilman 1997; Dessauer et al. 1999). This may be true for all CHDs. P-site inhibitors that, in the presence of pyrophosphate, are uncompetitive—dead-end inhibitors of the mammalian ACs—have been shown to bind to the VC_1-IIC_2 heterodimer active site, suggesting that these complexes may mimic binding of the products in the active sites (Tesmer et al. 2000). The binding sites of these inhibitors are very similar to that described above for the substrate analogs. However, as the P-site inhibitors do not significantly inhibit most homodimeric ACs, it is possible that complexes with mAC CHDs may not accurately represent the product-bound state in all CHDs. There may be subtle variations in the mode of binding—and consequently release—of products by different CHDs. For example, based on a superposition of cAMP on P-site inhibitors bound in the active site of the VC_1-IIC_2 heterodimer, it was proposed that steric clashes between the cyclized phosphate moiety and the A-site metal as well as the $\beta2^B$-$\beta3^B$ loop Asn^B resulted in the preferential release of cAMP (Tesmer et al. 2000). However, a similar steric restriction is not observed when cAMP is superimposed on the AMPCPP-bound structure of Rv1900c (Sinha et al. 2005). It is possible that catalysis proceeds only upon formation of a more closed conformation than observed in this structure of Rv1900c, which might then lead to a steric clash similar to that proposed for the mammalian ACs. Alternately, it is possible that the $\beta2^B$-$\beta3^B$ loop Asn^B that appears to regulate binding of the purine ring in Rv1900c, also regulates the release of the product cAMP by a similar mechanism.

Binding and catalysis by GGDEF domains

The structure of PleD, a DGC from *C. crescentus*, in complex with the product c-diGMP provides clues to the active site of the GGDEF domains (Chan et al. 2004). Like the CHDs, the GGDEF domain active site is located at the C-termini of the strands constituting the core βααββαβ motif. Further, like the CHDs, dimerization appears to be a prerequisite for catalysis. However, there appears to be no further similarity in the mode by which CHDs and GGDEF domains bind nucleotide. While in the CHDs, dimerization is required to bind a single NTP or cNMP, dimerization of the GGDEF domains is required because a single GGDEF domain appears to provide all of the determinants to bind only one GTP or half of a c-diGMP molecule while a second GGDEF domain, related to the first by the twofold symmetry axis of the c-diGMP, provides the determinants to bind the other GTP or the other half of the c-diGMP molecule (Fig. 5b and 7).

There are no metals bound in the structure of the c-diGMP-bound PleD complex, yet the two invariant aspartates that coordinate metal in the CHDs, the β1 Asp and β2-β3 loop Asp, are conserved at equivalent structural locations in GGDEF domains as an aspartate, and either an aspartate or a glutamate, respectively (Fig. 2a–b). However, the orientation of each guanine nucleotide moiety of diGMP in the catalytic site of PleD is the reverse of that of ATP analogs bound to CHDs, such that the ribose phosphate moiety is extended over the β2-β3 loop but directed away from the β1 strand (Fig. 7). In this orientation, the α-phosphate is hydrogen-bonded to the backbone amide of the second glycine of the GGDEF motif, which maps to the β2-β3 loop. The ribose ring is stabilized by hydrogen bonds between the ribose 2′-hydroxyl and the amide of a conserved α1 asparagine, as well as between the ribose 3′-hydroxyl and the amino group of a conserved lysine, which is the second residue of α1. The guanine ring is placed over a hydrophobic patch formed by side chains from α1 and α2 and is stabilized by edge-to-face stacking interactions with a conserved phenylalanine that is the first residue of α1 (Figs. 1 and 7). Hydrogen bonds involving the guanine N3/N2 and N1 with a conserved α1 Asn and α2 Asp, respectively, provide specificity for GTP over ATP.

Given the conservation of residues involved in key contacts with the guanine ring of c-diGMP, it is likely that the guanine ring from the substrate GTP also occupies a similar

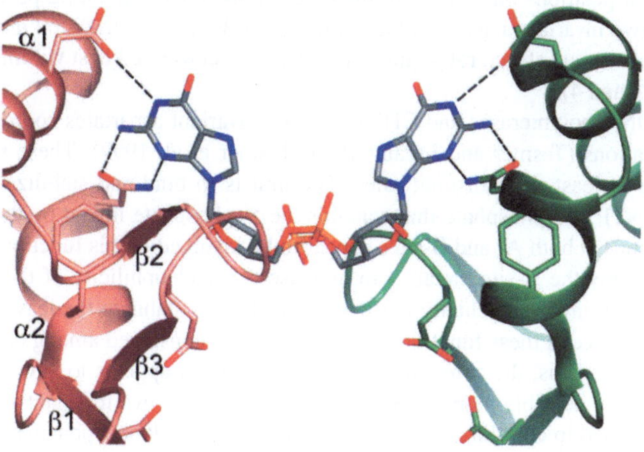

Fig. 7 Proposed active site of class III DGCs. The two GGDEF domains are colored as in Fig. 5 and secondary structures labeled as in Fig. 1. Molecular details of bound ligands and of protein residues that play a role in substrate binding and catalysis are shown. *Dashed, black lines* indicate hydrogen bonds

position and makes the same contacts. However, the position of the ribose and phosphate groups is likely to be flexible. Based on the c-diGMP-bound PleD structure, Chan et al. suggest that the mechanism of base recognition by the GGDEF domain sterically precludes binding of the GTP γ-phosphate over the β1-α1 loop, which constitutes the P-loop of CHDs. Instead, they predict that the β- and γ-phosphates of GTP are probably accommodated close to the first glycine of the GGDEF motif on the β2-β3 loop while a Mg^{2+} ion is coordinated by carboxylates from the glutamate constituting the fourth residue of this motif. Such a binding mode for GTP would preclude involvement of a second metal ion coordinated by the β1 aspartate and the β2-β3 loop glutamate/aspartate, which are conserved in both CHDs and GGDEF domains, as this metal would be too distant from the ribose 3′-hydroxyl or pyrophosphate group to play any role in binding or catalysis. Thus, the proposed mechanism employs only a single metal ion bound at a site distinct from the A- and B-metal sites of CHD dimers. Chan et al. further suggest that the central acidic residue of the GGDEF motif may act as a general base to deprotonate the GTP 3′-hydroxyl, priming it for attack on the 5′-phosphate of another GTP molecule. Dimers of GTP-bound GGDEF domains would orient the activated GTP molecules such that the 3′-hydroxyl of one is poised to attack the 5′-phosphate of the other. In this scheme, the amino group of the conserved α1 Lys would serve to stabilize the pentavalent phosphoryl transition state, as well as the pyrophosphate leaving group. This mechanism of catalysis does not define the role of the conserved β1 Asp that is invariant among GGDEF domains, CHDs, and DNA polymerases. Thus, although the c-diGMP-bound PleD structure has provided valuable insights into the structure and mode of product binding by GGDEF domains, structures of GGDEF domains in complex with substrate analogs and metal ions will be crucial to defining the mode of binding substrate and the mechanism of catalysis employed by DGCs.

Evolution of class III NC active sites

It is highly probable that CHDs, GGDEF domains, and palm domains of type I DNA polymerases all evolved from a common ancestor. These domains not only share a common core βααββαβ structural motif (Fig. 2), but also two invariant acidic residues located at topologically equivalent positions on this motif, the C-terminus of β1 and the β2-β3 loop (Fig. 3). Further, these two invariant acidic residues, and the βααββαβ core structural motif, also constitute the sequence and structural features, respectively, conserved best within each type of domain (Figs. 1 and 4).

In type I DNA polymerases and CHDs, the two invariant aspartates coordinate two divalent metal cations (Tesmer and Sprang 1998; Tesmer et al. 1999). These metal cations appear to have at least three distinct roles. The first is to bind and stabilize the negative charge on the NTP polyphosphate that contains the 5′-phosphate linked by the phosphodiester bond. Metals at both A- and B-sites appear to be involved in this function. The second role, involving only the A-site metal, is to increase the nucleophilicity of the 3′-hydroxyl. And the third role, involving chiefly the B-site metal, is to stabilize the leaving pyrophosphate group. All three of these functions appear to be well-conserved among both CHDs and DNA polymerases. Thus, the core βααββαβ structural motif appears to have evolved from a more ancient, and possibly more promiscuous, motif bearing two invariant acidic residues optimally positioned to coordinate two metals to bind NTP polyphosphates and to catalyze the formation of 3′-5′-phosphodiester bonds. The role of the equivalent acidic residues of GGDEF domains and any metals they may coordinate remains to be verified. In the GGDEF domains, an additional glutamate, corresponding to the fourth residue of the GGDEF motif, is conserved at the N-terminus of β3 and has been postulated to coordinate metal at a site distinct from either the A- or B-metal sites (Chan et al. 2004). Interestingly, although not present in the CHDs, a glutamate is also highly conserved at an equivalent location in the type I DNA polymerases, which, like the GGDEF domains, catalyze the formation of intermolecular 3′-5′-phosphodiester bonds.

Glycines on the β2-β3 loop and β4 are invariant in almost all class III NCs, indicating that these residues arose early in the evolution of these NCs, probably concurrently with the class III NC fold. The absence of a side-chain at the invariant β2-β3 loop glycine position prevents occlusion of purine-binding subsite in the CHDs, and probably part of the nucleotide-binding site in GGDEF domains as well. Additionally, in CHDs, the Φ and Ψ dihedral angles of this glycine map to unfavorable regions of the Ramachandran plot, facilitating a specific conformation of the β2-β3 loop and allowing one of the peptide planes of this glycine to stabilize the ATP purine ring by stacking interactions. Thus, this glycine plays a direct role in binding substrate in CHDs. The β4 glycine facilitates packing of α4 against β4, thereby positioning β4 relative to the βααββαβ motif. In CHDs, α4 contributes important residues to the active sites of these MNCs.

Although the βααββαβ structural motif and the two invariant acidic residues constitute a conserved module used to bind polyphosphate groups and catalyze the formation of 3′-5′-phosphodiester bonds by the class III MNCs and DGCs as well as the type I DNA polymerases, each enzyme family has evolved markedly different elements to bind and recognize the purine and ribose groups of the substrate NTPs. In type I DNA polymerases, the elements that interact with the nucleotide base and ribose are chiefly provided by the fingers domain, which is inserted between α1 and α2 of the βααββαβ structural motif. The fingers domains are typically much more divergent than the βααββαβ structural motif-containing palm domain of these enzymes (Brautigam and Steitz 1998). In CHDs, elements from the second structural module, comprising the class III NC and CHD-specific insertions, are responsible for several of the key interactions with the nucleotide base and ribose. However, in contrast to DNA-polymerases and CHDs, the c-diGMP bound PleD structure suggests that in GGDEF domains elements of the βααββαβ structural motif may also be responsible for binding purine and ribose.

While DNA polymerases function as monomers to catalyze the formation of 3′-5′-phosphodiester bonds, the class III NCs appear to be active only as dimers. However, the GGDEF domains and CHDs are expected to have markedly different modes of dimerization. Two GGDEF domains are expected to associate to form a homodimer with a single, interfacial, catalytically competent active site. The two subunits of a GGDEF domain dimer are expected to be related by the dyad symmetry of the bound substrates or products, with each subunit providing identical determinants to bind a single GTP molecule or half a c-diGMP molecule. In contrast, two CHDs associate to form two potential active sites at the dimer interface. As discussed in section "CHD active sites", each of these potential active sites is lined by elements from two structural modules, the first module comprising the βααββαβ structural motif of one subunit and the second module comprising the class III NC and CHD-specific insertions of the other subunit. In CHDs, therefore, elements of the second structural module—especially the CHD-specific insertions, the mode of dimerization, and of binding NTPs—all probably evolved concurrently from a more primitive βααββαβ structural motif.

CHDs from lower organisms appear to function as homodimers with two chemically identical, interfacial active sites related by the dimer dyad axis. In most of these CHD homodimers the two dyad-related active sites are also structurally and mechanistically equivalent and symmetrical (Fig. 8a; Steegborn et al. 2005b). Unexpectedly, however, although the Rv1900c CHDs form homodimers with chemically equivalent, dyad-related active sites,

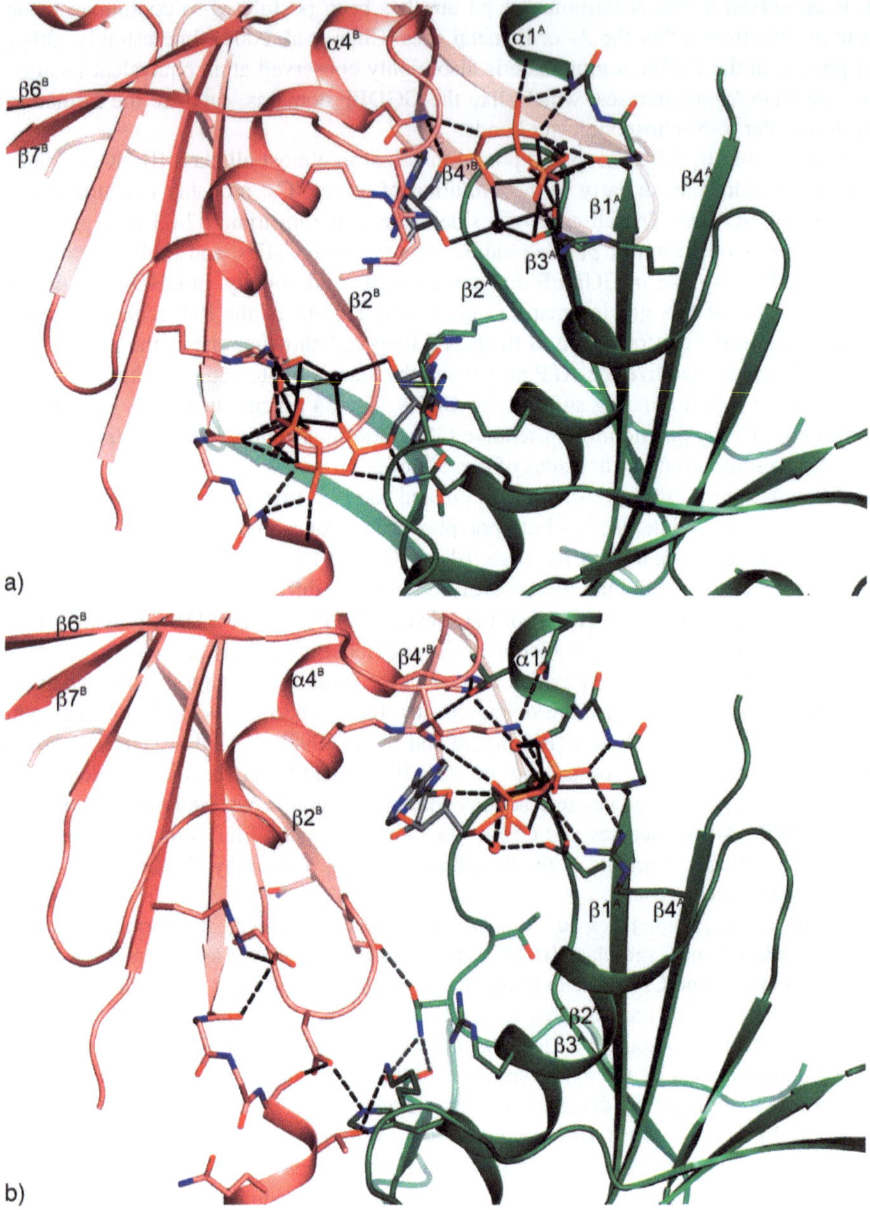
β6B
β7B
α4B
α1A
β4'B
β4A
β1A
β3A
β2B
β2A
a)
β6B
β7B
α4B
β4'B
α1A
β2B
β1A
β4A
β2A
β3A
b)

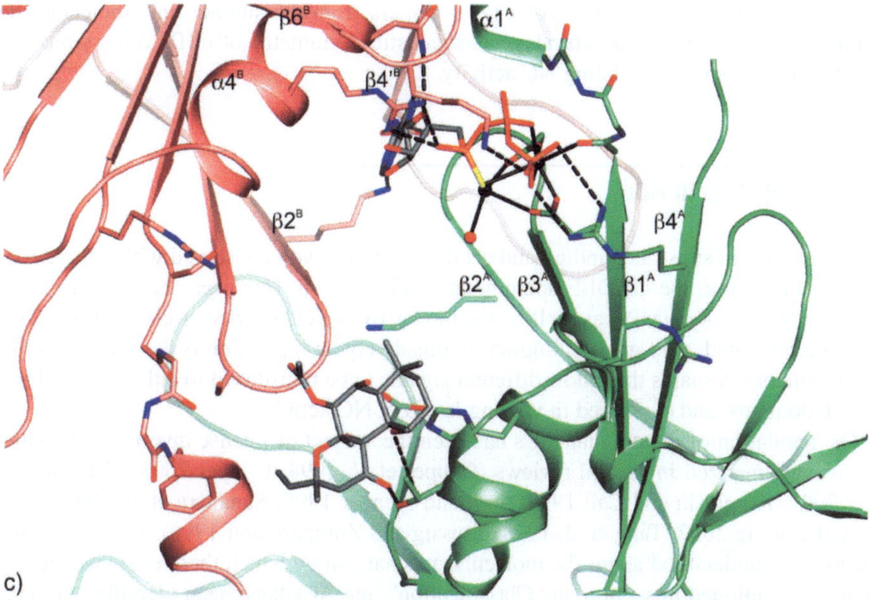

Fig. 8a–c Dyad-related sites of CHD dimers. Subunits, molecular details, bonds, and labels are depicted as in Fig. 6. **a** The two symmetrical, dyad-related active sites of the *S. platensis* CyaC CHD homodimer. Both sites are occupied by ATPαS. **b** The two chemically equivalent, yet structurally and mechanistically asymmetrical, dyad-related active sites of the Rv1900c CHD homodimer. Only one active site is occupied by AMPCPP. **c** The two asymmetrical, dyad-related binding sites of the VC_1-IIC_2 heterodimer include a single functional active site occupied by ATPαS and a nonfunctional site occupied by forskolin

these sites are structurally asymmetrical (Fig. 8b; Sinha et al. 2005). This asymmetry, which is amplified in the AMPCPP-bound, catalytically competent state, appears to be a consequence of the β2-β3 hairpin that participates simultaneously in both active sites of the dimer. As discussed earlier, this hairpin bears a conserved polar-X-X-G-D motif, of which the first residue participates in one active site, while the last two residues contribute to the dyad-related active site. The first residue of this motif interacts with the purine base, the glycine is important for structure and also for stabilizing the bound purine ring, and the aspartate is essential for binding metal and polyphosphate, and for catalysis. While the first polar residue of this motif is an adenine-specifying lysine or guanine-specifying glutamate in ACs and GCs, respectively, in Rv1900c it is an asparagine. Instead of hydrogen-bonding to the purine ring, this asparagine from one CHD points away from the AMPCPP-bound site, while in the other it appears to partially obstruct the ATP purine-binding subsite, preventing ATP from binding in this site (Fig. 8b; Sinha et al. 2005). Thus, although Rv1900c is catalytically active as a homodimer, the two active sites of the homodimer appear incapable of simultaneously binding ATP and catalyzing the synthesis of cAMP, suggesting a mechanism of half-of-sites reactivity. Mutational and biochemical data provide further support for a mechanism of half-of-sites reactivity (Sinha et al. 2005). Finally, while the two potential active sites in homodimeric ACs are chemically identical, those of the pseudo-symmetrical CHD heterodimers are not. Thus, in contrast to the homodimeric ACs, usually only one of the two potential active sites of the pseudo-symmetrical mammalian CHD heterodimers has the full complement of residues required for catalysis (Fig. 8c; Tesmer et al. 1997). The mechanism of half-of-sites reactivity observed for homodimeric Rv1900c may provide clues to the mechanistic variability that may have preceded and facilitated the evolution of the more complex, asymmetrical, intramolecular heterodimers from the simpler, intermolecular homodimers. This chemical, structural, or mechanistic asymmetry of CHD dimers provides a unique mechanism of regulating NC activity.

Regulation of class III NCs

The mechanism of substrate binding and catalysis is fairly well conserved within the CHDs, and probably within the GGDEF domain subfamily as well. However, the stimuli that affect the synthesis of cNMPs, as well as the modes by which activity of class III MNCs or DGCs are regulated, vary greatly. Implicit in this divergent regulation is the presence of diverse regulatory domains that allow different signals to be transduced to either the CHDs or GGDEF domains, and expressed in varying levels of NC activity.

The regulation of mammalian ACs has been the subject of intense investigation, which has been summarized in several reviews (Cooper et al. 1994; Cooper 2003; Hanoune and Defer 2001; Krupinski and Cali 1998; Smit and Iyengar 1998; Sunahara et al. 1996; Sunahara and Taussig 2002; Tang et al. 1998; Taussig and Zimmermann 1998). However, much remains to be understood about the molecular mechanism by which these enzymes are regulated. As mentioned in section on "Classification", ten ACs have been identified in mammals. Mammalian ACs have a fairly uniform architecture. The CHDs of all ten mammalian ACs function as intramolecular heterodimers. Nine of these ACs are mACs, while the tenth is a "soluble" AC (sAC) (Buck et al. 1999). There are several striking differences between the mACs and sAC. The nine mACs are homologs consisting of two consecutive repeats of a hexa-helical transmembrane domain and a cytoplasmic CHD. The sAC gene contains two tandem cytoplasmic CHDs linked to a "AAA" nucleotide binding domain (NBD) followed by a segment of approximately 800 residues that is uncharacterized in structure and function and the sAC appears to be active as various C-terminally truncated forms (Buck et al. 1999; Feng et al. 2005; Geng et al. 2005; Jaiswal and Conti 2001; Roelofs and Van Haastert 2002).

The role of the two hexa-helical transmembrane domains, M_1 and M_2, of the mACs is not well understood. It is likely that as a membrane anchor for the CHDs, the transmembrane domains not only serve to localize the cytoplasmic, catalytic CHDs to membrane surfaces but also are essential for targeting to the plasma membrane, where they are in close proximity to various effector molecules that regulate the mACs. Further, it is likely that these domains play key roles in AC oligomerization. Experiments using various combinations of truncated, duplicated, inverted, chimeric, or fully swapped membrane anchors from different mAC homologs indicated that the M_1 and M_2 domains also form intramolecular heterodimers and that type-specific association is critical for enzyme activity (Seebacher et al. 2001). These type-specific interactions may be important mechanisms for preventing inappropriate cross-isoform interactions. Further, the M_2 domains may also be involved in intermolecular homodimerization, which may be indicative of mechanisms of inhibition involving the formation of catalytically incompetent CHD homodimers (Gu et al. 2002). It has also been suggested that the transmembrane domains may enable mACs to specifically associate with other membrane proteins, and these interactions balanced against homo-oligomerization properties of mACs may play an important role in regulation of catalytic activity. Thus, different types of mACs may localize to separate membrane rafts along with the specific integral-transmembrane proteins or membrane-anchored proteins that regulate them, creating local microdomains of finely tuned cAMP levels within a single cell (Cooper 2003, 2005; Noyama and Maekawa 2003; Zehmer and Hazel 2003). Finally, some studies suggest that the transmembrane domains may also serve as direct sensors of membrane capacitance and potential, or membrane lipid and fluidity composition and regulate CHD activity in response to these stimuli (Cooper et al. 1998; Reddy et al. 1995b).

The mACs are differentially regulated by numerous effectors including the G proteins that transduce the numerous signals sensed by G protein-coupled receptors; Ca^{2+}, that regulates ACs either directly or indirectly via calmodulin; kinases such as calmodulin kinase protein kinase C and protein kinase A; and phosphatases such as calcineurin (Cooper et al. 1994, 1995; Cooper 2003; Hanoune and Defer 2001; Krupinski and Cali 1998; Mons et al. 1998; Smit and Iyengar 1998; Sunahara et al. 1996; Sunahara and Taussig 2002; Tang et al. 1998; Taussig and Zimmermann 1998). Unlike the mACs, the sAC does not appear to be regulated by any of these macromolecular effectors, but instead appears to be directly regulated by small molecules such as Ca^{2+}, bicarbonate, and catechols (Cann et al. 2003; Chen et al. 2000; Geng et al. 2005; Hyne and Garbers 1979; Jaiswal and Conti 2003; Litvin et al. 2003). The role of the NBDs and the transmembrane domain in these ACs is not yet understood.

Compared to higher eukaryotes, signaling pathways that regulate MNCs in prokaryotes and lower eukaryotes are simpler and more direct. Indeed, most CHDs from these lower organisms are linked in *cis* to one or more of a large variety of sensory/regulatory domains that are thought to directly transduce signals to the CHDs. Homologs of many of these domains have been shown to play similar regulatory roles in other signal transduction pathways. These domains include:

- Small molecule-binding domains such as GAF (an abbreviation derived from domains in cGMP-binding diesterases, adenylyl cyclases, and the *E. coli* transcription factor, FhlA); PAS (domains in period clock protein, Aryl hydrocarbon receptor, single-minded protein); and BLUF (flavin-containing, PAS domain like blue-light receptor) domains. Domains with diverse functions such as HAMP (domains in histidine kinases, adenylyl cyclases, methyl-accepting chemotaxis proteins and phosphatases) domains; ATPase domains; histidine kinase domains; RAS-associated domains; two-iron, two-sulfur cluster containing domains; leucine-rich repeats; TPR (tetratricopeptide repeat) domains; helix-turn-helix DNA binding domains; single helical transmembrane anchors; hexa-helical transmembrane domains. A few domains whose fold and function have not yet been identified (Linder 2005; Shenoy and Visweswariah 2004).

In addition, CHDs from lower organisms may be directly regulated by small molecules such as bicarbonate and Ca^{2+} (Cann et al. 2003; Chen et al. 2000; Masuda and Ono 2005; Steegborn et al. 2005b). Little is known about the molecular mechanism by which these diverse regulators influence CHD activity, although recent studies have begun to investigate the structural basis of these mechanisms.

Structures of the mammalian VC_1-IIC_2 heterodimer in complex with the cAMP synthesis-stimulating Gα subunit ($G\alpha_s$) greatly enhanced our understanding of how the mACs are regulated by macromolecular effectors (Tesmer et al. 1997, 1999, 2000). The most extensive interaction between $G\alpha_s$ and the VC_1-IIC_2 heterodimer involves the insertion of the $G\alpha_s$ switch II helix into a groove between the $\alpha1^B$-$\alpha2^B$ and $\alpha3^B$-$\beta4^B$ loops of IIC_2 (Fig. 9a). As $G\alpha_s$ can bind to IIC_2 monomers and disrupt IIC_2 homodimers in the absence of VC_1, it is likely that this interaction provides most of the binding energy required for the formation the AC-$G\alpha_s$ complex and simultaneously prevents formation of nonproductive C_2 homodimers. Although homodimerization of C_1 or C_2 domains would be irrelevant to regulation in vivo for monomeric mAC molecules, it could be a significant mechanism of regulation if mACs assemble as dimers or oligomers (Gu 2002). In addition to the switch II-mediated interaction, the α3-β5 loop of $G\alpha_s$ interacts with the N-terminus of VC_1, which packs against the CHD dimerization arm^A (Tesmer et al. 1997). Although this interaction is not essential for binding of $G\alpha_s$ to mACs, it is essential for the upregulation of catalytic activity by $G\alpha_s$ (Sunahara et al. 1997). This interaction may promote the formation of a more closed conformation of the intramolecular C_1-C_2 heterodimers essential for activity. Indeed, binding of potent substrate analogs, such as ATPαS, is accompanied by movement of C_1 toward the C_2 domain, thereby aligning the P-loop^A of C_1 with the purine-binding elements in $\beta4'^B$ and $\alpha4^B$ of C_2, and closing the catalytic site (Fig. 9b; Tesmer et al. 1999). This transition may be abetted by the binding of $G\alpha_s$. Finally, mutation data suggest that other regions of $G\alpha_s$, such as the α4-β6 loop, may be involved in additional interactions, presumably with AC regions not included in the VC_1 and IIC_2 constructs used for the crystal studies (Scholich et al. 1997b; Sunahara et al. 1997; Yan et al. 1997a). Thus, $G\alpha_s$ functions as an allosteric activator of mammalian ACs by facilitating both the relative rearrangement of the C_1 and C_2 domains to form catalytically competent heterodimers, as well as by stabilizing catalytically competent conformations of loops and residues involved in catalysis.

$G\alpha_s$-mediated stimulation of type I, V, and VI mACs is opposed by the inhibitory Gα subunit, $G\alpha_i$ (Chen et al. 1997; Taussig et al. 1993). Myristoylation of the $G\alpha_i$ subunit is required for this inhibitory activity. $G\alpha_i$ does not compete for the AC binding site of the $G\alpha_s$ subunit; rather, the inhibitory and stimulatory Gα subunits appear to bind to distinct sites on mACs (Taussig et al. 1994; Wittpoth et al. 1999). Evidence that $G\alpha_i$ and VC_1 form a 1:1 complex, combined with mutational analysis, suggest that switch II, and perhaps the α4-β6 loop of $G\alpha_i$, are involved in the interaction with VC_1, while the $\alpha1^A$-$\alpha2^A$ and $\alpha3^A$-$\beta4^A$ loops of VC_1 play key roles in binding of $G\alpha_i$ (Dessauer et al. 1998; Wittpoth et al. 1999). Thus, the homologous G_α subunits appear to exert opposite regulatory effects by binding to pseudo dyad-related sites on the intramolecular, heterodimeric mammalian C_1-C_2 heterodimers (Fig. 9a). Experiments with extended constructs indicate that regions outside of the C_1 CHD domain are also involved in the interaction with $G\alpha_i$ (Dessauer et al. 1998). As $G\alpha_i$ is a noncompetitive inhibitor of $G\alpha_s$ -stimulated AC, it is probable that $G\alpha_i$ binds directly to the $G\alpha_s$-bound AC complex. The structural mechanism of inhibition is unknown. However, it has not been possible to isolate a ternary complex between $G\alpha_i$, $G\alpha_s$, and isolated C_1 and C_2 domains (Dessauer et al. 1998). Therefore it is possible that $G\alpha_i$ weakens the C_1-C_2 interface and prevents essential conformational changes in regions key to catalysis. Since C_1-C_2 heterodimers have been captured only in $G\alpha_s$- and forskolin-activated states, the basal, ligand-free state of the C_1-C_2 heterodimers, and consequently, the changes wrought by their regulators, cannot be accurately defined. Further, it is probable that, like $G\alpha_i$, other regulators of mACs such as Gβγ, calmodulin kinase II, and protein kinase C, that act chiefly on the $G\alpha_s$-stimulated state of ACs, also function by regulating conformational changes at the domain interface. Unlike the pseudo dyad-related binding sites of the

Fig. 9 Regulation and conformational change in the mammalian mACs. **a** Asymmetric regulation of the VC_1-IIC_2 heterodimer by $G\alpha_s$. VC_1, IIC_2, and $G\alpha_s$ are colored *salmon*, *green*, and *violet*, respectively. Molecular details of the substrate analog ATPαS and the activator forskolin bound to the VC_1-IIC_2 heterodimer and the GTP analog bound to $G\alpha_s$ are shown. Selected structural elements of VC_1-IIC_2 and the switch II helix of $G\alpha_s$ are labeled. **b** Conformational change in the $G\alpha_s$ and forskolin-activated VC_1-IIC_2 heterodimer upon binding of substrate. VC_1 and IIC_2 subunits are colored *khaki* and *gray*, respectively, in the "open" state crystallized in the absence of substrate, and *green* and *salmon*, respectively, in the presence of ATPαS; $G\alpha_s$, forskolin, and ATPαS are not shown. Secondary structure elements, β1-α1-α2 segment of VC_1 and β6-β7 hairpin of IIC_2, which show the largest displacements upon binding of ATPαS, are indicated ▶

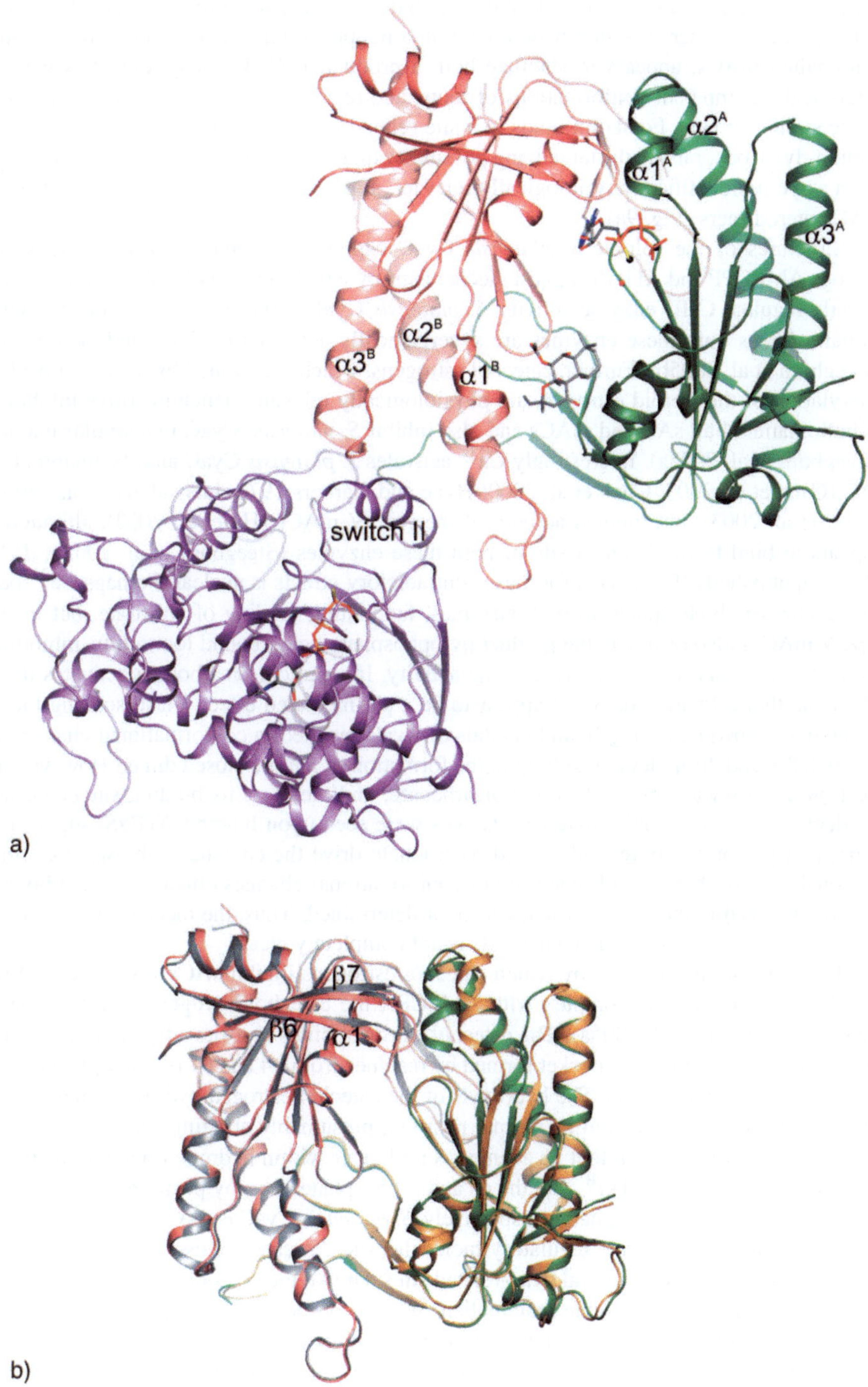
α2^A
α1^A
α3^A
α2^B
α3^B
α1^B
switch II
a)
β7
β6
α1
b)

homologous $G\alpha_s$ and $G\alpha_i$ subunits, these regulators are expected to bind asymmetrically to C_1-C_2 heterodimer. The diterpene forskolin, a nonphysiological activator of most of the mammalian mACs, appears to stabilize both a productive C_1-C_2 heterodimer as well as catalytically competent conformations of loops and residues involved in catalysis, playing a role similar to $G\alpha_s$. Forskolin binds to a site related to the heterodimer active site by the pseudo-dyad axis, raising the tantalizing possibility that this might constitute the binding site of an as-yet-unidentified, physiologically relevant, small-molecule regulator of mammalian mAC heterodimers (Fig. 9a).

Structures of the soluble *S. platensis* CyaC captured in complex with the substrate analogs AMPCPP and ATPαS suggest mechanisms by which small-molecule regulators may directly regulate CHD enzyme activity. *S. platensis* CyaC serves as a model for the mammalian sAC as both these enzymes are synergistically activated by Ca^{2+} and bicarbonate (Steegborn et al. 2005b). Further, catechol estrogens, which are formed by the oxidative hydroxylation of the steroid estrogen, are physiologically relevant, noncompetitive inhibitors of both mammalian sAC and mACs and also inhibit *S. platensis* CyaC in a similar manner (Steegborn et al. 2005a). Interestingly Ca^{2+} activates *S. platensis* CyaC and the mammalian sAC (Cann et al. 2003; Chen et al. 2000; Hyne and Garbers 1979; Jaiswal and Conti 2003; Litvin et al. 2003), but inhibits activity of the type V mACs (Hu et al. 2002), although it appears to bind to the B-metal site of both these enzymes (Steegborn et al. 2005b; T.-C. Mou, unpublished). The reason for these contradictory effects is unclear. Perhaps Ca^{2+} better stabilizes polyphosphate in both enzymes, facilitating binding of substrate, but in the type V mAC it also results in the product pyrophosphate being bound too tightly, inhibiting product release and consequently enzyme activity. In solution, bicarbonate activates these sACs and this activation does not appear to be a pH-mediated effect. Flash-soaking bicarbonate into substrate analog-bound crystals of CyaC resulted in conformational changes in the α1-helix and β6-β7 loop, resulting in the formation of a more closed dimer. However, no electron density was observed for the bicarbonate, and therefore its binding site could not be identified. Similar conformational changes were seen upon binding ATPαS, suggesting either that potent substrate analogs and bicarbonate drive the enzyme to the same conformational state or that the full repertoire of conformational changes effected by bicarbonate are not observed in the crystal structures so far determined. Thus, the mechanism by which bicarbonate and Ca^{2+} alone activate CyaC is not completely clear.

The probable mechanism by which catechol estrogens inhibit sACs was suggested by the structure of CyaC in complex with AMPCPP, the catechol estrogens, and two metal cations (Steegborn et al. 2005a). The catechol estrogen binds at a site adjacent to the CHD active site, in a hydrophobic pocket formed by residues from α4 and β1 of one CHD and the β2-β3 hairpin of both CHDs. The hydroxyls of the catechol estrogen hydroxyls chelate and displace the A-site metal from its normal position, presumably aborting catalysis. Further, AMPCPP is displaced such that its adenine is no longer within hydrogen-bonding distance of the dimerization arm^B Thr^B, and the α- and β-phosphates occupy positions typically occupied by the β- and γ-phosphates, respectively. *S. platensis* CyaC is a symmetrical, homodimeric enzyme (Fig. 8a), and regulatory molecules such as catechol estrogen and Ca^{2+} are identically bound to symmetrical dyad-related sites. Unlike the homodimer CyaC, however, the tandem CHDs of the soluble mammalian AC are not identical in sequence, and consequently the catalytic dimer formed by these domains must be pseudosymmetrical. Substitution of one of the two conserved catalytic aspartate residues in the C_1 domain by cysteine, and an insertion of several residues in the $\beta2^A$-$\beta3^A$ hairpin suggests that mammalian sAC possesses only one functional catalytic site. It is therefore possible that the asymmetry of the mammalian sAC will also be reflected in the mechanism of its regulation, reminiscent of the mammalian mACs.

M. tuberculosis Rv1900c is a two-domain, homodimeric MNC. The N-terminal domain bears sequence similarity to members of the α/β hydrolase family of lipases but has no catalytic activity toward known substrates, and appears to have weak, negative regulatory activity (Sinha et al. 2005). The C-terminal CHD, upon homodimerization, can use both ATP and GTP as substrate but has a 14-fold preference for ATP. Structures of Rv1900c CHD homodimers with and without the substrate analog AMPCPP demonstrated that even in the absence of additional regulatory stimuli, binding of substrate induces conformational changes within each CHD accompanied by a dramatic rotation of 16.6° and translation of 11 Å of the CHDs relative to each other, to form a catalytically competent dimer (Fig. 10a). As discussed previously, the Rv1900c CHD homodimers are structurally and mechanistically asymmetrical in both open and closed states. The role of the Rv1900c N-terminal regulatory domain in the regulation of the transition between inactive, open and the active, closed dimers, its influence on the structural and mechanistic asymmetry observed in the isolated Rv1900c CHD homodimers, and its effect upon substrate specificity has not yet been established.

M. tuberculosis Rv1264 is a two-domain AC comprising an N-terminal, 10-helix regulatory domain and a C-terminal CHD that also functions as a homodimer (Linder et al. 2002). Rv1264 is a pH-regulated enzyme that presumably plays a role in mycobacterial survival in the acidic environment of the phagolysosome. In the activated state, and even in the absence of substrate, the two CHDs form a homodimer whose subunits are arranged in a closed conformation similar to that seen in the ligand-bound structures of other CHDs (Fig. 10b; Tews et al. 2005). This active homodimer contains two identical, symmetrical active sites, each of which contains a complete, appropriately located complement of residues required for

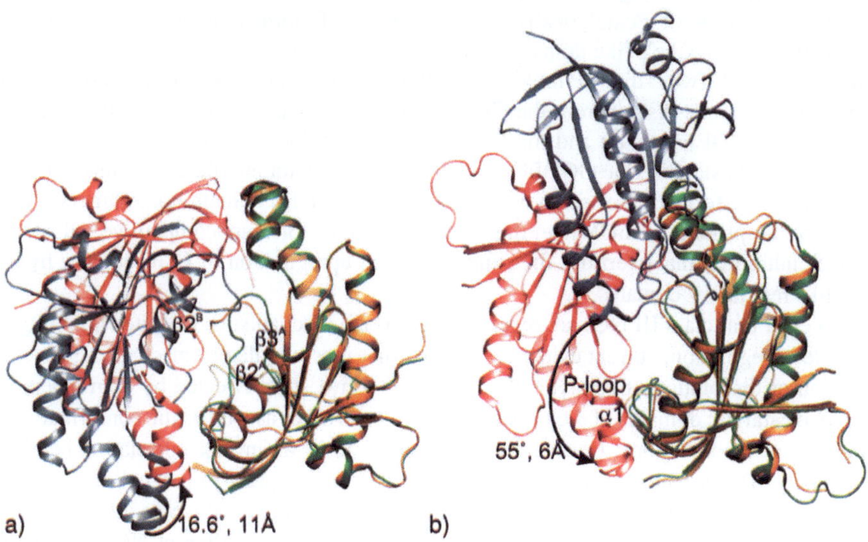

Fig. 10a, b Conformational change in two mycobacterial NCs. CHDs are colored *gray* and *khaki* in the open, catalytically incompetent conformation, and *salmon* and *green* in the closed, catalytically competent, conformation. The movement of the CHDs constituting a dimer, relative to each other, as well as the secondary structure elements that undergo the largest conformational changes are indicated. **a** Conformational change in Rv1900c CHD homodimers upon binding of substrate. The change from an open to a closed state is triggered by binding of the ATP analog, AMPCPP. **b** Conformational change in Rv1264 CHD dimers upon activation by pH

catalysis. Structures of the Rv1264 holoprotein crystallized at neutral and acidic pH demonstrate that CHD activity is also regulated by dramatic conformational changes, wherein the α1-helix of the CHD unfolds, destroying the ATP phosphate-binding subsite; concomitantly, the active sites of the dimer are disassembled by a dramatic rotation of 55° and translation of 6 Å of the two CHDs relative to each other (Fig. 10b; Tews et al. 2005). The regulatory domain of each subunit of the homodimer makes extensive contacts with the CHD of the other in both active and inactive states, and the pH-triggered inactivating transformation is effected by the pH-dependent extension of the C-terminal helix of the regulatory domain by four turns (Fig. 11). Despite these large conformational changes, in both active and inactive states, Rv1264 forms symmetrical homodimers and the regulatory domains from each of the two subunits that constitute the homodimer make identical, symmetrical contacts with the trans CHD (Fig. 11).

Like the prokaryotic CHDs, GGDEF domains appear to be covalently linked to a wide range of sensory/regulatory domains, but very little is known about the mechanism by which these domains regulate DGC activity. However, it appears that like the CHDs, regulating the formation of catalytically competent GGDEF domain dimers may be a key mode of regulation in DGCs. *C. crescentus* PleD, the only DGC of known structure, contains a CheY-like receiver domain a second CheY-like adaptor domain, in addition to the GGDEF domain (Chan et al. 2004). The structure of the nonphosphorylated form of the molecule provides clues to the mechanism by which activity of PleD is regulated (Chan et al. 2004). Inactivate PleD appears to be a monomer in solution. The sensor histidine kinase phosphorylates a receiver domain aspartate located at the interface of the receiver and adaptor domains, which is predicted to induce repacking of this interface and subsequent reorientation of the receiver and adaptor domains relative to each other. In this activated orientation, the receiver domain of one DGC can form isologous contacts with the adaptor domain of another activated DGC, enhancing formation of a catalytically competent GGDEF domain homodimer. In addition to regulation by the CheY-like domains, PleD is subject to tight feedback inhibition by the product c-diGMP. Two c-diGMP molecules with mutually intercalated purine bases bind as a single unit at a site distant from the GGDEF domain active site, located at the interface between the GGDEF domain and the CheY-like adaptor domain. Binding at this site has been proposed to stabilize the GGDEF domain-adaptor domain interface in an inactive state that hampers the formation of catalytically competent GGDEF domain homodimers. Thus, binding of the product, c-diGMP, to this inhibitory site enables noncompetitive, allosteric, feedback inhibition that overrides the stimulus-driven upregulation of DGC activity by the CheY-like regulatory domains.

It is clear that class III NCs are regulated by diverse mechanisms. Despite variability in the mode of regulation, current evidence suggests that most regulatory mechanisms involve formation or dissolution of catalytically competent active sites, effected by rearrangement of the two catalytic domains of the dimer relative to each other as well as the conformational changes in active site loops. In general, there appears to be an inverse correlation between the diversity in the architecture of the class III NCs and the complexity of signaling pathways that interact with them. Thus, NCs from lower organisms that are active as intermolecular homodimers show greater variation in architecture, enabling the different NCs to respond directly to different stimuli. In contrast, MNCs from higher organisms have a more uniform architecture, possibly having evolved by gene duplication of a primitive CHD to form intramolecular CHD heterodimers. The emergence of more complex regulatory networks accompanied the divergence of the duplicated genes and the subsequent pseudo-symmetry inherent in the heterodimers. These features have enabled MNCs to serve as integration points for networks of signaling pathways.

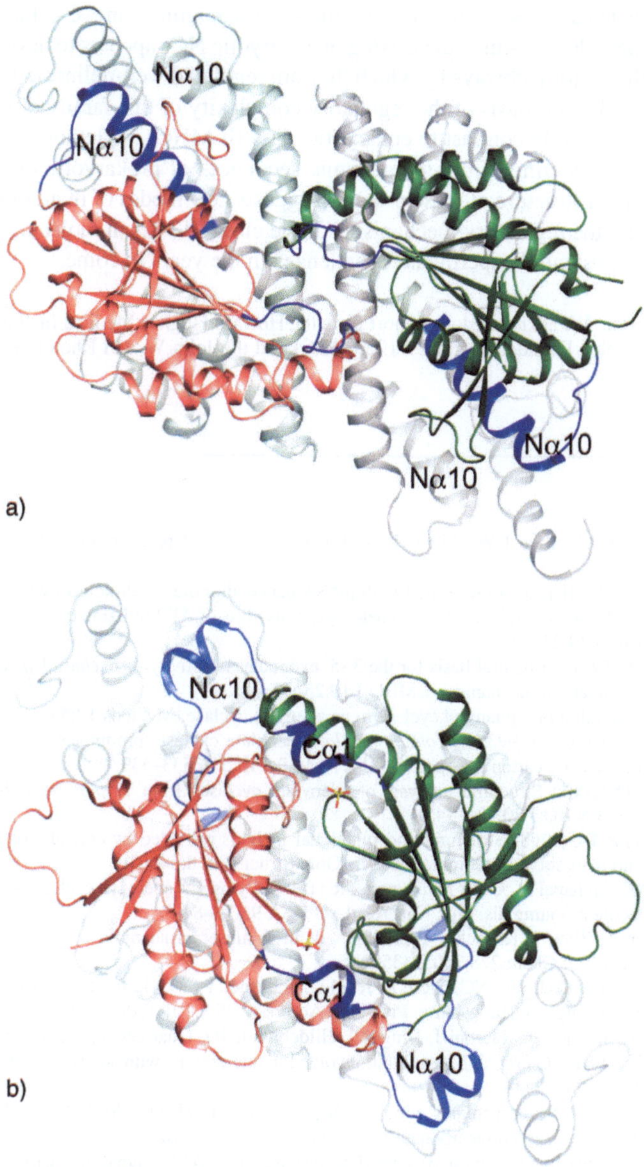

Fig. 11a, b pH-triggered conformational change and regulation of Rv1264. The subunits are colored *green* and *salmon*, with the regulatory domain of each subunit overlaid with *gray*. Secondary structure elements that undergo dramatic conformational changes are labeled and highlighted in *blue*. **a** The open, catalytically incompetent conformation. The C-terminal helix (*Nα10*) of each N-terminal, regulatory domain is elongated by four turns (colored *blue*). The P-loop and the α1 helix of each CHD (also colored *blue*) are unfolded. **b** The closed, catalytically competent conformation. The last four turns of Nα10 unfold and are rearranged. The P-loop and the α1 helix (labeled Cα1, colored *blue*) of each CHD fold into conformations similar to that seen in other CHD structures. A sulfate molecule bound in the polyphosphate-binding site is depicted in molecular detail

Nearly a decade has passed since the structures of the mammalian AC catalytic modules were first published, during which the catalog of eukaryotic NCs appears to have been nearly completed and the major pathways by which they are regulated well-delineated. However, it is clear that the molecular basis of the regulatory complexity of this family of enzymes has yet to be fully explored. In contrast to eukaryotic class III MNCs, a plethora of prokaryotic enzymes have been identified from the genomic sequences of prokaryotic organisms, and the divergent domain structure of these proteins suggests new modes of regulation. Thus, efforts to understand the regulatory mechanisms of the class III NCs in molecular detail shall continue to pose substantial experimental challenges in the years to come.

Acknowledgements. We acknowledge support from the Howard Hughes Medical Institute, NIH grant DK46371 (SRS), Welch Foundation grant I-1229 (SRS) and the John W. and Rhonda K. Pate Professorship to SRS.

References

Artymiuk PJ, Poirette AR, Rice DW, Willett P (1997) A polymerase I palm in adenylyl cyclase? Nature 388:33–34

Barton GJ (1993) ALSCRIPT: a tool to format multiple sequence alignments. Protein Eng 6:37–40

Barzu O, Danchin A (1994) Adenylyl cyclases: a heterogeneous class of ATP-utilizing enzymes. Prog Nucleic Acid Res Mol Biol 49:241–283

Beese LS, Steitz TA (1991) Structural basis for the 3′-5′ exonuclease activity of *Escherichia coli* DNA polymerase I: a two metal ion mechanism. EMBO J 10:25–33

Beuve A (1999) Conversion of a guanylyl cyclase to an adenylyl cyclase. Methods 19:545–550

Beuve A, Krin E, Danchin A (1993) *Rhizobium meliloti* adenylate cyclase: probing of a NTP-binding site common to cyclases and cation transporters. C R Acad Sci III 316:533–539

Bieger B, Essen LO (2001) Structural analysis of adenylate cyclases from *Trypanosoma brucei* in their monomeric state. EMBO J 20:433–445

Brautigam CA, Steitz TA (1998) Structural and functional insights provided by crystal structures of DNA polymerases and their substrate complexes. Curr Opin Struct Biol 8:54–63

Buck J, Sinclair ML, Schapal L, Cann MJ, Levin LR (1999) Cytosolic adenylyl cyclase defines a unique signaling molecule in mammals. Proc Natl Acad Sci USA 96:79–84

Cann MJ, Hammer A, Zhou J, Kanacher T (2003) A defined subset of adenylyl cyclases is regulated by bicarbonate ion. J Biol Chem 278:35033–35038

Chan C, Paul R, Samoray D, Amiot NC, Giese B, Jenal U, Schirmer T (2004) Structural basis of activity and allosteric control of diguanylate cyclase. Proc Natl Acad Sci USA 101:17084–17089

Chen Y, Weng G, Li J, Harry A, Pieroni J, Dingus J, Hildebrandt JD, Guarnieri F, Weinstein H, Iyengar R (1997) A surface on the G protein beta-subunit involved in interactions with adenylyl cyclases. Proc Natl Acad Sci USA 94:2711–2714

Chen Y, Cann MJ, Litvin TN, Iourgenko V, Sinclair ML, Levin LR, Buck J (2000) Soluble adenylyl cyclase as an evolutionarily conserved bicarbonate sensor. Science 289:625–628

Cooper D (2005) Compartmentalization of adenylate cyclase and cAMP signalling. Biochem Soc Trans 33:1319–1322

Cooper DM, Mons N, Fagan K (1994) Ca(2+)-sensitive adenylyl cyclases. Cell Signal 6:823–840

Cooper DM, Mons N, Karpen JW (1995) Adenylyl cyclases and the interaction between calcium and cAMP signalling. Nature 374:421–424

Cooper DM, Schell MJ, Thorn P, Irvine RF (1998) Regulation of adenylyl cyclase by membrane potential. J Biol Chem 273:27703–27707

Cooper DMF (2003) Regulation and organization of adenylyl cyclases and cAMP. Biochem J 375:517–529

Cotta M, Whitehead T, Wheeler M (1998) Identification of a novel adenylate cyclase in the ruminal anaerobe, *Prevotella ruminicola* D31d. FEMS Microbiol Lett 164:257–260

Coudart-Cavalli MP, Sismeiro O, Danchin A (1997) Bifunctional structure of two adenylyl cyclases from the myxobacterium *Stigmatella aurantiaca*. Biochimie 79:757–767

Dessauer CW, Gilman AG (1997) The catalytic mechanism of mammalian adenylyl cyclase. Equilibrium binding and kinetic analysis of P-site inhibition. J Biol Chem 272:27787–27795

Dessauer CW, Scully TT, Gilman AG (1997) Interactions of forskolin and ATP with the cytosolic domains of mammalian adenylyl cyclase. J Biol Chem 272:22272–22277

Dessauer CW, Tesmer JJ, Sprang SR, Gilman AG (1998) Identification of a Gia binding site on type V adenylyl cyclase. J Biol Chem 273:25831–25839

Dessauer CW, Tesmer JJ, Sprang SR, Gilman AG (1999) The interactions of adenylate cyclases with P-site inhibitors. Trends Pharmacol Sci 20:205–210

Doublié S, Ellenberger T (1998) The mechanism of action of T7 DNA polymerase. Curr Opin Struct Biol 8:704–712

Drum CL, Yan SZ, Bard J, Shen YQ, Lu D, Soelaiman S, Grabarek Z, Bohm A, Tang WJ (2002) Structural basis for the activation of anthrax adenylyl cyclase exotoxin by calmodulin. Nature 415:396–402

Eckstein F, Romaniuk PJ, Heideman W, Storm DR (1981) Stereochemistry of the mammalian adenylate cyclase reaction. J Biol Chem 256:9118–9120

Feng Q, Zhang Y, Li Y, Liu Z, Zuo J, Fang F (2005) Two domains are critical for the nuclear localization of soluble adenylyl cyclase. Biochimie 88:319–328

Gallagher DT, Smith NN, Kim SK, Heroux A, Robinson H, Reddy PT (2006) Structure of the Class IV Adenylyl Cyclase Reveals a Novel Fold. J Mol Biol. 2006 Aug 11; [Epub ahead of print]

Geng W, Wang Z, Zhang J, Reed BY, Pak CYC, Moe OW (2005) Cloning and characterization of the human soluble adenylyl cyclase. Am J Physiol Cell Physiol 288:C1305–1316

Gu C, Cali JJ, Cooper DM (2002) Dimerization of mammalian adenylate cyclases. Eur J Biochem 269:413–421

Guo Q, Shen Y, Zhukovskaya NL, Florian J, Tang WJ (2004) Structural and kinetic analyses of the interaction of anthrax adenylyl cyclase toxin with reaction products cAMP and pyrophosphate. J Biol Chem 279:29427–29435

Guo Q, Shen Y, Lee YS, Gibbs CS, Mrksich M, Tang WJ (2005) Structural basis for the interaction of *Bordetella pertussis* adenylyl cyclase toxin with calmodulin. EMBO J 24:3190–3201

Guo YL, Seebacher T, Kurz U, Linder JU, Schultz JE (2001) Adenylyl cyclase Rv1625c of *Mycobacterium tuberculosis*: a progenitor of mammalian adenylyl cyclases. EMBO J 20:3667–3675

Hanoune J, Defer N (2001) Regulation and role of adenylyl cyclase isoforms. Annu Rev Pharmacol Toxicol 41:145–174

Holland MM, Leib TK, Gerlt JA (1988) Isolation and characterization of a small catalytic domain released from the adenylate cyclase from *Escherichia coli* by digestion with trypsin. J Biol Chem 263:14661–14668

Hu B, Nakata H, Gu C, De Beer T, Cooper DM (2002) A critical interplay between Ca^{2+} inhibition and activation by Mg^{2+} of AC5 revealed by mutants and chimeric constructs. J Biol Chem 277:33139–33147

Hurley J (1998) The adenylyl and guanylyl cyclase superfamily. Curr Opin Struct Biol 8:770–777

Hurley JH (1999) Structure, mechanism, and regulation of mammalian adenylyl cyclase. J Biol Chem 274:7599–7602

Hyne RV, Garbers DL (1979) Regulation of guinea pig sperm adenylate cyclase by calcium. Biol Reprod 21:1135–1142

Jaiswal BS, Conti M (2001) Identification and functional analysis of splice variants of the germ cell soluble adenylyl cyclase. J Biol Chem 276:31698–31708

Jaiswal BS, Conti M (2003) Calcium regulation of the soluble adenylyl cyclase expressed in mammalian spermatozoa. Proc Natl Acad Sci USA 100:10676–10681

Kasahara M, Unno T, Yashiro K, Ohmori M (2001) CyaG, a novel cyanobacterial adenylyl cyclase and a possible ancestor of mammalian guanylyl cyclases. J Biol Chem 276:10564–10569

Ketkar A, Shenoya A, Ramagopal UA, Visweswariaha SS, Sugun K (2006) A structural basis for the role of nucleotide specifying residues in regulating the oligomerization of the Rv1625c adenylyl cyclase from M. tuberculosis. J Mol Biol 356:904–916

Ketkar AD, Shenoy AR, Kesavulu MM, Visweswariah SS, Suguna K (2004) Purification, crystallization and preliminary X-ray diffraction analysis of the catalytic domain of adenylyl cyclase Rv1625c from *Mycobacterium tuberculosis*. Acta Crystallogr D Biol Crystallogr 60:371–373

Kimura Y, Vassylyev DG, Matsushima M, Mitsuoka K, Murata K, Hiral T, Fujiyoshi Y (1997) Surface of bacteriorhodopsin revealed by high-resolution electron crystallography. Nature 389:206–211

Krupinski J, Cali JJ (1998) Molecular diversity of the adenylyl cyclases. Adv Second Messenger Phosphoprotein Res 32:53–79

Ladant D, Ullmann A (1999) *Bordetella pertussis* adenylate cyclase: a toxin with multiple talents. Trends Microbiol 7:172–176

Leppla SH (1982) Anthrax toxin edema factor: a bacterial adenylate cyclase that increases cyclic AMP concentrations of eukaryotic cells. Proc Natl Acad Sci USA 79:3162–3166

Linder JU (2005) Substrate selection by class III adenylyl cyclases and guanylyl cyclases. IUBMB Life 57:797–803

Linder JU, Schultz JE (2003) The class III adenylyl cyclases: multi-purpose signalling modules. Cell Signal 15:1081–1089

Linder JU, Engel P, Reimer A, Kruger T, Plattner H, Schultz A, Schultz JE (1999) Guanylyl cyclases with the topology of mammalian adenylyl cyclases and an N-terminal P-type ATPase-like domain in Paramecium, Tetrahymena and Plasmodium. EMBO J 18:4222–4232

Linder JU, Hoffmann T, Kurz U, Schultz JE (2000) A guanylyl cyclase from Paramecium with 22 transmembrane spans. Expression of the catalytic domains and formation of chimeras with the catalytic domains of mammalian adenylyl cyclases. J Biol Chem 275:11235–11240

Linder JU, Schultz A, Schultz JE (2002) Adenylyl cyclase Rv1264 from *Mycobacterium tuberculosis* has an autoinhibitory N-terminal domain. J Biol Chem 277:15271–15276

Linder JU, Hammer A, Schultz JE (2004) The effect of HAMP domains on class IIIb adenylyl cyclases from *Mycobacterium tuberculosis*. Eur J Biochem 271:2446–2451

Litvin NT, Kamenetsky M, Zarifyan A, Buck J, Levin LR (2003) Kinetic properties of "soluble" adenylyl cyclase. Synergism between calcium and bicarbonate. J Biol Chem 278:15922–15926

Liu Y, Ruoho AE, Rao VD, Hurley JH (1997) Catalytic mechanism of the adenylyl and guanylyl cyclases: modeling and mutational analysis. Proc Natl Acad Sci USA 94:13414–13419

Masuda S, Ono TA (2005) Adenylyl cyclase activity of Cya1 from the cyanobacterium *Synechocystis* sp. strain PCC 6803 is inhibited by bicarbonate. J Bacteriol 187:5032–5035

McCue LA, McDonough KA, Lawrence CE (2000) Functional classification of cNMP-binding proteins and nucleotide cyclases with implications for novel regulatory pathways in *Mycobacterium tuberculosis*. Genome Res 10:204–219

Mons N, Decorte L, Jaffard R, Cooper DM (1998) Ca^{2+}-sensitive adenylyl cyclases, key integrators of cellular signalling. Life Sci 62:1647–1652

Mou TC, Gille A, Fancy DA, Seifert R, Sprang SR (2005) Structural basis for the inhibition of mammalian membrane adenylyl cyclase by 2′(3′)-O-(N-methylanthraniloyl)-guanosine 5′-triphosphate. J Biol Chem 280:7253–7261

Murzin AG (1998) How far divergent evolution goes in proteins. Curr Opin Struct Biol 8:380–387

Noyama K, Maekawa S (2003) Localization of cyclic nucleotide phosphodiesterase 2 in the brain-derived Triton-insoluble low-density fraction (raft). Neurosci Res 45:141–148

Ochoa de Alda JAG, Ajlani G, Houmard J (2000) *Synechocystis* strain PCC 6803 cya2, a prokaryotic gene that encodes a guanylyl cyclase. J Bacteriol 182:3839–3842

Pei J, Grishin N (2001) GGDEF domain is homologous to adenylyl cyclase. Proteins 42:210–216

Reddy P, Hoskins J, McKenney K (1995a) Mapping domains in proteins: dissection and expression of *Escherichia coli* adenylyl cyclase. Anal Biochem 231:282–286

Reddy R, Smith D, Wayman G, Wu Z, Villacres EC, Storm DR (1995b) Voltage-sensitive adenylyl cyclase activity in cultured neurons. A calcium-independent phenomenon. J Biol Chem 270:14340–14346

Roelofs J, Van Haastert PJM (2002) Deducing the origin of soluble adenylyl cyclase, a gene lost in multiple lineages. Mol Biol Evol 19:2239–2246

Romling U, Gomelsky M, Galperin MY (2005) C-di-GMP: the dawning of a novel bacterial signalling system. Mol Microbiol 57:629–639

Scholich K, Barbier AJ, Mullenix JB, Patel TB (1997a) Characterization of soluble forms of nonchimeric type V adenylyl cyclase. Proc Natl Acad Sci USA 94:2915–2920

Scholich K, Wittpoth C, Barbier AJ, Mullenix JB, Patel TB (1997b) Identification of an intramolecular interaction between small regions in type V adenylyl cyclase that influences stimulation of enzyme activity by Gsalpha. Proc Natl Acad Sci USA 94:9602–9607

Seebacher T, Linder JU, Schultz JE (2001) An isoform-specific interaction of the membrane anchors affects mammalian adenylyl cyclase type V activity. Eur J Biochem 268:105–110

Shen Y, Zhukovskaya NL, Guo Q, Florián J, Tang WJ (2005) Calcium-independent calmodulin binding and two-metal-ion catalytic mechanism of anthrax edema factor. EMBO J 24:929–941

Shenoy A, Visweswariah S (2004) Class III nucleotide cyclases in bacteria and archaebacteria: lineage-specific expansion of adenylyl cyclases and a dearth of guanylyl cyclases. FEBS Lett 561:11–21

Shenoy AR, Srinivasan N, Subramaniam M, Visweswariah SS (2003) Mutational analysis of the *Mycobacterium tuberculosis* Rv1625c adenylyl cyclase: residues that confer nucleotide specificity contribute to dimerization. FEBS Lett 545:253–259

Shenoy AR, Sreenath NP, Mahalingam M, Visweswariah SS (2005) Characterization of phylogenetically distant members of the adenylate cyclase family from mycobacteria: Rv1647 from *Mycobacterium tuberculosis* and its orthologue ML1399 from M. leprae. Biochem J 387:541–551

Simonds WF (1999) G protein regulation of adenylate cyclase. Trends Pharmacol Sci 20:66–73

Sinha SC, Wetterer M, Sprang SR, Schultz JE, Linder JU (2005) Origin of asymmetry in adenylyl cyclases: structures of *Mycobacterium tuberculosis* Rv1900c. EMBO J 24:663–673

Sismeiro O, Trotot P, Biville F, Vivares C, Danchin A (1998) *Aeromonas hydrophila* adenylyl cyclase2: a new class of adenylyl cyclases with thermophilic properties and sequence similarities to proteins from hyperthermophilic archaebacteria. J Bacteriol 180:3339–3344

Smit MJ, Iyengar R (1998) Mammalian adenylyl cyclases. Adv Second Messenger Phosphoprotein Res 32:1–21

Steegborn C, Litvin TN, Hess KC, Capper AB, Taussig R, Buck J, Levin LR, Wu H (2005a) A novel mechanism for adenylyl cyclase inhibition from the crystal structure of its complex with catechol estrogen. J Biol Chem 280:31754–31759

Steegborn C, Litvin TN, Levin LR, Buck J, Wu H (2005b) Bicarbonate activation of adenylyl cyclase via promotion of catalytic active site closure and metal recruitment. Nat Struct Mol Biol 12:32–37

Steitz TA (1993) DNA- and RNA-dependent DNA polymerases. Curr Opin Struct Biol 3:31–38

Steitz TA (1999) DNA polymerases: structural diversity and common mechanisms. J Biol Chem 274:17395–17398

Steitz TA, Steitz JA (1993) A general two-metal-ion mechanism for catalytic RNA. Proc Natl Acad Sci USA 90:6498–6502

Steitz TA, Smerdon SJ, Jäger J, Joyce CM (1994) A unified polymerase mechanism for nonhomologous DNA and RNA polymerases. Science 266:2022–2025

Süsstrunk U, Pidoux J, Taubert S, Ullmann A, CJ T (1998) Pleiotropic effects of cAMP on germination, antibiotic biosynthesis and morphological development in *Streptomyces coelicolor*. Mol Microbiol 30:33–46

Sunahara RK, Taussig R (2002) Isoforms of mammalian adenylyl cyclase: multiplicities of signaling. Mol Interv 2:168–184

Sunahara RK, Dessauer CW, Gilman AG (1996) Complexity and diversity of mammalian adenylyl cyclases. Annu Rev Pharmacol Toxicol 36:461–480

Sunahara RK, Dessauer CW, Whisnant RE, Kleuss C, Gilman AG (1997) Interaction of Gsa with the cytosolic domains of mammalian adenylyl cyclase. J Biol Chem 272:22265–22271

Sunahara RK, Beuve A, Tesmer JJG, Sprang SR, Garbers DL, Gilman AG (1998) Exchange of substrate and inhibitor specificities between adenylyl and guanylyl cyclases. J Biol Chem 273:16332–16338

Tang WJ, Gilman AG (1995) Construction of a soluble adenylyl cyclase activated by Gs alpha and forskolin. Science 268:1769–1772

Tang WJ, Stanzel M, Gilman AG (1995) Truncation and alanine-scanning mutants of type I adenylyl cyclase. Biochemistry 34:14563–14572

Tang WJ, Yan S, Drum CL (1998) Class III adenylyl cyclases: regulation and underlying mechanisms. Adv Second Messenger Phosphoprotein Res 32:137–151

Taussig R, Zimmermann G (1998) Type-specific regulation of mammalian adenylyl cyclases by G protein pathways. Adv Second Messenger Phosphoprotein Res 32:81–98

Taussig R, Iniguez-Lluhi JA, Gilman AG (1993) Inhibition of adenylyl cyclase by Gi alpha. Science 261:218–221

Taussig R, Tang WJ, Hepler JR, Gilman AG (1994) Distinct patterns of bidirectional regulation of mammalian adenylyl cyclases. J Biol Chem 269:6093–6100

Tellez-Sosa J, Soberon N, Vega-Segura A, Torres-Marquez ME, Cevallos MA (2002) The rhizobium etli cyaC product: characterization of a novel adenylate cyclase class. J Bacteriol 184:3560–3568

Tesmer JJ, Sprang SR (1998) The structure, catalytic mechanism and regulation of adenylyl cyclase. Curr Opin Struct Biol 8:713–719

Tesmer JJ, Dessauer CW, Sunahara RK, Murray LD, Johnson RA, Gilman AG, Sprang SR (2000) Molecular basis for P-site inhibition of adenylyl cyclase. Biochemistry 39:14464–14471

Tesmer JJG, Sunahara RK, Gilman AG, Sprang SR (1997) Crystal structure of the catalytic domains of adenylyl cyclase in a complex with Gsa_GTPgS. Science 278:1907–1916

Tesmer JJG, Sunahara RK, Johnson RA, Gilman AG, Sprang SR (1999) Two metal ion catalysis in adenylyl cyclase. Science 285:756–760

Tews I, Findeisen F, Sinning I, Schultz A, Schultz JE, Linder JU (2005) The structure of a pH-sensing mycobacterial adenylyl cyclase holoenzyme. Science 308:1020–1023

Tucker CL, Hurley JH, Miller TR, Hurley JB (1998) Two amino acid substitutions convert a guanylyl cyclase, RetGC-1, into an adenylyl cyclase. Proc Natl Acad Sci USA 95:5993–5997

Weber JH, Vishnyakov A, Hambach K, Schultz A, Schultz JE, Linder JU (2004) Adenylyl cyclases from plasmodium, paramecium and tetrahymena are novel ion channel/enzyme fusion proteins. Cell Signal 16:115–125

Whisnant RE, Gilman AG, Dessauer CW (1996) Interaction of the two cytosolic domains of mammalian adenylyl cyclase. Proc Natl Acad Sci USA 93:6621–6625

Wittpoth C, Scholich K, Yigzaw Y, Stringfield TM, Patel TB (1999) Regions on adenylyl cyclase that are necessary for inhibition of activity by beta gamma and G(ialpha) subunits of heterotrimeric G proteins. Proc Natl Acad Sci USA 96:9551–9556

Yahr TL, Vallis AJ, Hancock MK, Barbieri JT, Frank DW (1998) ExoY, an adenylate cyclase secreted by the *Pseudomonas aeruginosa* type III system. Proc Natl Acad Sci USA 95:13899–13904

Yan SZ, Hahn D, Huang ZH, Tang WJ (1996) Two cytoplasmic domains of mammalian adenylyl cyclase form a Gsa- and forskolin-activated enzyme in vitro. J Biol Chem 271:10941–10945

Yan SZ, Huang ZH, Rao VD, Hurley JH, Tang WJ (1997a) Three discrete regions of mammalian adenylyl cyclase form a site for Gsalpha activation. J Biol Chem 272:18849–18854

Yan SZ, Huang ZH, Shaw RS, Tang WJ (1997b) The conserved asparagine and arginine are essential for catalysis of mammalian adenylyl cyclase. J Biol Chem 272:12342–12349

Zehmer JK, Hazel JR (2003) Plasma membrane rafts of rainbow trout are subject to thermal acclimation. J Exp Biol 206:1657–1667

Zhang G, Liu Y, Ruoho AE, Hurley JH (1997) Structure of the adenylyl cyclase catalytic core. Nature 386:247–253

Zimmermann G, Zhou D, Taussig R (1998) Mutations uncover a role for two magnesium ions in the catalytic mechanism of adenylyl cyclase. J Biol Chem 273:19650–19655

MIX
Papier aus verantwortungsvollen Quellen
Paper from responsible sources
FSC® C105338

If you have any concerns about our products,
you can contact us on
ProductSafety@springernature.com

In case Publisher is established outside the EU,
the EU authorized representative is:
Springer Nature Customer Service Center GmbH
Europaplatz 3, 69115 Heidelberg, Germany

Printed by Libri Plureos GmbH
in Hamburg, Germany